REMINISCENCES OF THE VIENNA CIRCLE AND THE MATHEMATICAL COLLOQUIUM

VIENNA CIRCLE COLLECTION

VOLUME 20

VOLUME EDITOR: BRIAN McGUINNESS

KARL MENGER

REMINISCENCES OF THE VIENNA CIRCLE AND THE MATHEMATICAL COLLOQUIUM

Edited by

LOUISE GOLLAND, BRIAN McGUINNESS and ABE SKLAR

KLUWER ACADEMIC PUBLISHERS
DORDRECHT / BOSTON / LONDON

Library of Congress Cataloging-in-Publication Data

Menger, Karl, 1902–
    Reminiscences of the Vienna Circle and the Mathematical Colloquium
/ by Karl Menger ; edited by Louise Golland, Brian McGuinness, Abe
Sklar.
        p.   cm. -- (Vienna Circle collection ; v. 20)
    Includes index.
    ISBN 0-7923-2711-X
    1. Mathematics--Austria--Vienna--History--20th century. 2. Vienna
circle. 3. Menger, Karl, 1902–   . I. Golland, Louise, 1942–   .
II. McGuinness, Brian. III. Sklar, A. IV. Title. V. Series.
QA27.A9M46 1994
510'.9436'1309042--dc20   94-5014

ISBN 0-7923-2711-X (hb)
ISBN 0-7923-2873-6 (pb)

Published by Kluwer Academic Publishers,
P.O. Box 17, 3300 AA Dordrecht, The Netherlands.

Kluwer Academic Publishers incorporates
the publishing programmes of
D. Reidel, Martinus Nijhoff, Dr W. Junk and MTP Press.

Sold and distributed in the U.S.A. and Canada
by Kluwer Academic Publishers,
101 Philip Drive, Norwell, MA 02061, U.S.A.

In all other countries, sold and distributed
by Kluwer Academic Publishers Group,
P.O. Box 322, 3300 AH Dordrecht, The Netherlands.

Printed on acid-free paper

Printed in the Netherlands

TABLE OF CONTENTS

Karl Menger
(1902-1985)
Bibliopolis (Napels), 1987.
Reprinted with permission.

INTRODUCTION

Karl Menger was born in Vienna on January 13, 1902, the only child of two gifted parents. His mother Hermione, née Andermann (1870-1922), in addition to her musical abilities, wrote and published short stories and novelettes, while his father Carl (1840-1921) was the noted Austrian economist, one of the founders of marginal utility theory. A highly cultured man, and a liberal rationalist in the nineteenth century sense, the elder Menger had witnessed the defeat and humiliation of the old Austrian empire by Bismarck's Prussia, and the subsequent establishment under Prussian leadership of a militaristic, mystically nationalistic, state-capitalist German empire — in effect, the first modern "military-industrial complex."

These events helped frame in him a set of attitudes that he later transmitted to his son, and which included an appreciation of cultural attainments and tolerance and respect for cultural differences, combined with a deep suspicion of rabid nationalism, particularly the German variety. Also a fascination with structure, whether artistic, scientific, philosophical, or theological, but a rejection of any aura of mysticism or mumbo-jumbo accompanying such structure. Thus the son remarked at least once that the archangels' chant that begins the Prolog im Himmel in Goethe's *Faust* was perhaps the most beautiful thing in the German language "but of course it doesn't mean anything."

By the time the young Karl Menger entered the University of Vienna in the fall of 1920, he had absorbed these attitudes and discovered in himself an aptitude for science. His original intention was to study physics, and he did in fact attend lectures by Hans Thirring. It was not long, however, before he, at first tentatively then definitely, switched to mathematics. Much later, he jokingly remarked that the occasion for the definite abandonment of a possible career in physics was his unsuccessful attempt to read Hermann Weyl's classic book on relativity *Raum-Zeit-Materie*. More to the point was his discovery that he could make, and in fact had made, an important contribution to mathematics.

The story is quite dramatic, and Menger himself has told it in the chapter, "Why the Circle Invited Me. The Theory of Curves and Dimension Theory". Here is a capsule version: In the spring of 1921 he went to a lecture by the mathematician Hans Hahn, newly arrived in Vienna, on "What's new concerning the concept of a curve". In the lecture, Hahn pointed out that there was as yet no mathematically satisfactory definition of a curve. Menger pondered the question for several days, came up with what seemed like a reasonable definition and took it to Hahn. It must be emphasized that at that time, and indeed until quite recently, it was unusual for an undergraduate student in Central Europe to call upon a professor. Hahn listened, was impressed, and encouraged his visitor to keep working on the definition and on the inextricably linked notion of the dimension of a geometric object in general.

Menger did so, even after he developed tuberculosis in May 1921. By that fall, it was necessary for him to enter a sanatorium in the mountains of Styria. Before doing so, he deposited, according to a footnote on p. 132 of George Temple's book *100 Years of Mathematics* (Springer, 1981) "a sealed envelope containing his definition of a curve and a suggestion of his concept of dimension . . . with the Vienna Academy of Sciences . . . The envelope was opened in 1926 and its contents published . . Meanwhile [he] had also published . . . papers [in] 1923, 1924, 1926 . . "

These dates would be irrelevant if they did not bear on a question of priority. It happened that, concurrently with Menger and quite independently, the brilliant young Russian mathematician Pavel S. Uryson (1898-1924; the name generally appears as Paul Urysohn in Western publications) had developed an entirely equivalent though technically more complicated, definition of dimension. To quote again from Temple, p. 132: "According to Alexandroff['s 1925 obituary], Urysohn constructed his theory of dimension during the years 1921-2 and during this period communicated his results to the Mathematical Society of Moscow. A preliminary notice was published in 1922 and a full account in 1925 . . . Menger's definition is undoubtedly simpler and more general than Urysohn's, and the question of priority is of minor importance." Urysohn tragically drowned while swimming off the coast of France; he will be remembered in mathematics not only for dimension theory, but for other fundamental contributions, including some such as "Urysohn's Lemma" and the "Urysohn Metrization Theorem" to which his name is inseparably attached.

After almost a year at the sanatorium, Menger emerged completely cured and came back to Vienna. In 1924 he received his Ph.D. from the University of Vienna, and in March 1925, he left for Amsterdam, where he was to remain as a docent at the university for two years. He worked with the eminent and idiosyncratic mathematician who had invited him there, L. E. J. Brouwer.

Despite his increasingly strained relations with Brouwer, the years in Amsterdam were fruitful ones for him. He continued his work in dimension and curve theory and gained insight into logic and alternative ways of looking at mathematics as a whole.

In 1927 he returned to the University of Vienna as Professor of Geometry. He was asked to join the discussion group which had been organized during his absence around the philosopher Moritz Schlick later known as the Vienna Circle. Significant for mathematics was the Mathematical Colloquium which he organized. There faculty, students, and visitors could report on their own work, discuss outstanding problems, and listen to reviews of recent publications. The eight volumes of the *Ergebnisse eines mathematischen Kolloquiums*, covering the academic years from 1928-29 through 1935-36, were edited by Menger, and various of Menger's students, including Georg Nöbeling, Kurt Gödel and Abraham Wald.

Menger capped his work on dimension with the book *Dimensionstheorie* (1928, Teubner). Fifty years later, J. Keesling wrote in the Bulletin of American Mathematical Society (v. 83, p. 203): "This book has historical value. It reveals at one and the same time the naiveté of the early investigators by modern standards and yet their remarkable perception of what the important results were and the future direction of the theory. Copies are difficult to obtain now, but it is worth the effort."

One of these "important results" is worth citing. It states that every n-dimensional separable metric space is homeomorphic (topologically equivalent) to part of a certain 'universal' n-dimensional space, which can in turn be realized as a compact set in $(2n + 1)$-dimensional Euclidean space. This was proved for $n = 1$ in the 1928 book, where a sketch of a proof for all other n was also given. (Complete proofs for all n were given by Georg Nöbeling and others in 1931.) The universal n-dimensional spaces are now known as "Menger universal spaces" or "Menger universal compacta." The universal 1-dimensional space is also called the "Menger universal curve" or, because of its appearance as a set in 3-dimensional space, the "Menger sponge" (B. Mandelbrot, *The Fractal Geometry of Nature*, 1982, Freeman, p. 144).

Like his work in dimension theory, Menger's work in curve theory culminated in a book: *Kurventheorie* (1932, Teubner; reprinted 1967, Chelsea). Chapter VI of the book is devoted to a "Fundamental Theorem" about "the order of a regular point" on a curve. This Theorem follows from an "n-Arc Theorem", whose proof depends on a very special case called the "n-Chain Theorem for Graphs." This last theorem, first stated and proved as a lemma in a 1927 paper on curve theory has since become widely known in graph theory as "Menger's Theorem." In 1980, Menger was asked to write about the history of the theorem. The resulting paper, "On the Origin of the n-Arc Theorem" appeared in vol. 5 of the *Journal of Graph Theory*. In an accompaning Editor's Note, Frank Harary called Menger's Theorem "one of the most important results in graph theory" and "*the* fundamental theorem on connectivity in graphs."

Other work during this period includes: the first applications of metric space notions in geometry (rather than analysis or topology), culminating in metric theories of betweenness, segments, and geodesics, and laying the groundwork for the subject now known as "distance geometry"; investigations in the foundations of projective geometry, on acount of which Menger is now considered one of the pioneers of lattice theory; various papers in logic and foundations some of the more significant of which can be found in Menger's *Selected Papers*, published as no. 10 in the present collection, and perhaps most remarkable, a little book on the application of simple mathematical notions to ethical problems. This book was originally published by Springer in 1934, and was translated into English and republished under the title *Morality, Decision, and Social Organization* in 1974 as no. 6 of the present collection.

Menger describes the deteriorating political situation in Austria and his motivation for writing his book on ethics in the present volume. With the death of Hahn by natural causes and the murder of Schlick the Circle came to a sad end. Thus apart from his work the bright spots in the picture for Menger were his marriage in 1935 to the young actuarial student Hilda Axamit, and the birth in 1936 of the couple's first child Karl, Jr.

In 1936 Menger attended the International Congress of Mathematics in Oslo and was elected its president. Here friends and associates urged Menger to leave Austria. Menger obtained a professorship at Notre Dame, and the family settled in South Bend, Indiana in 1937.

Unlike many refugee intellectuals, Menger liked the United States, and soon felt completely at home in his new country. Other ex-European mathematicians were offended at being required to teach such elementary subjects as trigonometry; Menger enjoyed teaching undergraduate courses. Indeed, to the end of his days he regarded teaching, when properly done, as not only pleasurable in itself, but a stimulus rather than a hindrance to creative research and scholarship.

Through the efforts of friends, he was able to get his former student Abraham Wald out of post-Anschluss Austria and into the United States. He arranged for Kurt Gödel to visit Notre Dame and tried in vain to bring him there permanently (see his last chapter). He founded the series of Notre Dame Mathematical Lectures, at least one of which became widely known: several generations of American mathematicians used Number 2 in the series, Emil Artin's "Galois Theory" as their text in the subject.

Menger also organized a Mathematical Colloquium at Notre Dame in the spirit of the one that had existed in Vienna, and founded the *Reports of a Mathematical Colloquium, Second Series*. Like the Viennese *Ergebnisse eines mathematischen Kolloquiums*, the *Reports* carried articles not only by Menger and his students and colleagues at Notre Dame, but also by visitors and other colleagues. The first issue of the *Reports* appeared in 1938, the eighth and final one in 1948.

In the meantime, the Menger family was growing. Twins, Rosemary and Fred, were born in 1937, and the last child, Eve, in 1942. Their father took great delight in his young children: years later, he would tell how he marveled at the weeks-old Eve answering her father's smiling face with a smile of her own.

South Bend is not very far from Chicago, where Rudolf Carnap, one of the original members of the Vienna Circle, had come in 1936. Carnap and Charles W. Morris had organized a discussion group, inevitably called the "Chicago Circle", which met irregularly on Saturday mornings at the University of Chicago. As often as he could Menger came in from South Bend and participated in the sessions of the Circle.

The one tangible accomplishment of the Chicago Circle was to get some of its participants to write, and the University of Chicago Press to publish, the first monographs in the series called the *International Encyclopedia of Unified Science*. Apart from this the Circle suffered an early series of blows from which, although it continued to meet in a desultory fashion until the 70's, it never fully recovered. The first of these was the departure of the noted linguist Leonard Bloomfield from the University of Chicago to become Sterling Professor of Linguistics at Yale. As C. F. Hochett, the editor of *A Leonard Bloomfield Anthology* (Indiana University Press, 1970) remarks on page 543, "Bloomfield's call to Yale in 1940 . . . came during the Hutchins and Adler period at the University of Chicago . . . The authorities at the University of Chicago made no effort whatsoever to keep him. . . " Bloomfield had based his discussion of "calculation" in his monograph *Linguistic Aspects of Science* (Volume 1, Number 4 of the International Encyclopedia of Unified Science) on Menger's article "The New Logic"(*Philosophy of Science*, 4, 1937). Menger, in turn, esteemed Bloomfield as one of the rare kind of non-mathematicians who "understand what mathematics is all about."

The next major and practically fatal blow to the Chicago Circle was the war, which in the United States began in 1941, and which disrupted academic life in general. Menger, for instance, found himself teaching calculus to large classes of students in the Navy's V-12 program. Although he enjoyed teaching, the time and effort involved in this and and in the continuation of his own research, together with the wartime restrictions on travel, made it impossible for him to get to Chicago with any regularity. Other participants experienced similar difficulties. Thus although the Circle survived the war, it never became as influential or widely known as its Viennese predecessor.

In the meantime Menger's interests were beginning to turn to areas with less immediate *éclat* than dimension theory or curve theory. Three of such areas are non-Euclidean (specifically, hyperbolic) geometry, probabilistic geometry, and the algebra of functions.

It had long been known that plane hyperbolic geometry could be developed within a restricted region of the ordinary Euclidean plane. What Menger saw was that this meant that hyperbolic geometry could be given an axiomatic foundation that was not only independent of, but essentially simpler than, any possible one for Euclidean geometry. This has philosophical as well as mathematical import. For everyone had agreed that Euclidean geometry is simpler than hyperbolic geometry (for example, in the hyperbolic plane there are 11 different kinds of conic sections as compared with just 3 different kinds in the Euclidean plane). Thus the great French mathematician Henri Poincaré could say at the end of Chapter 8 of his *Science and Hypothesis* (1905, republished by Dover 1952): "Now, Euclidean geometry is . . . the simplest in itself, just as a polynomial of the first degree is simpler than a polynomial of the second degree . . ." But by 1940, this "simplest" structure was seen to rest on a *necessarily* more complicated foundation, thus bringing the absolute character of the notion of "simplicity" into question.

Unlike the earlier work on dimension theory and curve theory, the work of Menger and his students on hyperbolic geometry made little stir. This can be explained by the general neglect of geometry which persisted until recently, and in particular, by the common feeling that hyperbolic geometry was a historically interesting amusing mathematical sideline, apart from a few technical applications in complex analysis. This attitude has changed, and there has been a great revival of interest in hyperbolic geometry which may lead to a renewed appreciation of the work of Menger and his students on the foundations of the subject.

For a long time, Menger thought of introducing probabilistic notions into geometry. The basic reason why one would want to do this is that geometry is in large part (though by no means as a whole) concerned with properties of structures — positions, distances, areas, volumes; etc. — that can be measured. Now measurement is always inexact in practice, and there are scientific theories such as quantum mechanics that imply that at least some measurements are necessarily inexact *in principle*. Such considerations led Menger to publish a brief note in the December 1942 *Proceedings of the National Academy of Sciences*. In this note, entitled "Statistical Metrics", he showed how one could replace a numerical distance between points p and q by a function F_{pq} whose value $F_{pq}(x)$ at the real number x is interpreted as the probability that the distance between p and q is less than x.

A. Wald read the note with great interest and in the following year published a note of his own on the subject in the same *Proceedings* (a slightly expanded version appears in his posthumous *Selected Papers*). Menger and Wald discussed the matter at length when they met during summer vacations, but their projected collaboration was aborted by Wald's death in 1950 in a plane crash in India. Profoundly moved by the death of his friend, Menger did little work on the subject for several years. Gradually, however, others have taken up the work, and beginning at the point where Menger and Wald left off, have built up a substantial theory. Those who have worked on what are now called "probabilistic metric spaces" unanimously feel that the profound significance of the subject begun by Karl Menger in 1942 will be increasingly recognized; it may be remarked that some of the auxiliary machinery of the theory has already been applied to areas as diverse as cluster analysis, measures of statistical dependence, hysteresis, and dose-response problems in medicine.

The third significant field initiated by Menger at Notre Dame was the Algebra of Functions. This began with an observation common to everyone who has taught a course in calculus; the functions dealt with in such a course are combined with one another in pairs by addition, by multiplication and by composition (also called "substitution"). Thus there are, for example, rules for differentiating a sum, a product, or a composite of two functions (the last usually appearing in somewhat disguised form as "the chain rule"). Menger not only made this observation, he acted upon it. He elucidated the algebraic structure that underlies these and other rules, and thus invented what he called "tri-operational algebra". He and his students developed this in a series of papers in the *Reports of a Mathematical Colloquium* and in Number 3 of the *Notre Dame Mathematical Lectures*, the one entitled "Algebra of Analysis".

He extended the scope of his work on the algebra of functions after moving, in 1948, to the Illinois Institute of Technology in Chicago, recently formed by the merger of two existing institutions. While continuing his study of 1-place functions ("functions of a single variable" in the usual terminology), he turned his attention towards multiple functions ("functions of several variables") where, he said in "The algebra of functions: past, present, future" [*Rendiconti di Matematica* 20, 1961, pp. 409-430], "the real future of the algebra of functions seems to me to lie . . .." Certain combinations of multiplace functions have a characteristic property which Menger called "superassociativity". The extensive investigations by Menger and his students and colleagues have led others, particularly Russian mathematicians, to label algebraic structures in which superassociativity holds "Menger algebras". Such structures have found application in several areas of mathematics, notably logic and, to Menger's delight, geometry.

One great disappointment of Menger's intellectual life was the failure of his long and strenuous effort to reform both the teaching and the application of calculus and other undergraduate mathematics. Like the work on the algebra of functions and in fact closely associated with it, the reform effort resulted from Menger's experience in teaching calculus classes in the United States. Many teachers, upon finding that their students meet difficulties, blame the students themselves for all such difficulties, and blithely go on in their accustomed way. Menger, feeling that such a response was inadequate and unjust, analysed all aspects of the situation, and concluded that the traditional way of presenting the calculus was deeply flawed. He resolved to remedy this situation. After some years of preparation and testing, he began to teach what he called his "modern approach" to calculus in 1952. Concurrently, he tried to influence the entire mathematical community by a series of articles and lectures. In them, and in the book he produced, initially in mimeographed form, for his class, he analyzed the manifold notions connected with the word "variable", and carefully distinguished arguments and values of functions from the functions themselves. The latter being the actual objects treated in calculus, he introduced explicit symbols for certain important functions (most notably, the identity function) which because of their ubiquity had been taken for granted and so had remained undesignated. A feature of the book which is often overlooked and which is perhaps its most striking pedagogical innovation is the development of a complete "miniature calculus" within which all the basic features of the calculus, including the so-called "Fundamental Theorem", can be illustrated without bringing in limits.

The few reviews the book received were generally favorable; some, such as the ones by Cain T. Adamson in the *Mathematical Gazette* and H. E. Bray in the *American Mathematical Monthly* (61, 1954, pp. 483-492), exceptionally so. Bray ended his review (of the second mimeographed edition) by saying: "Unfortunately, the typography of the book is unattractive, but this fault will undoubtedly be remedied by an enterprising publisher who will print the book in the attractive format in which so many lesser books are now being offered."

In 1955 the book did get published as *Calculus: A Modern Approach*, by Ginn and Company, though in a format perhaps more accurately described as "modest" than as "attractive". In this form it sold steadily but slowly; it has now been long out of print and is difficult to obtain.

There is not enough space to discuss the rest of Menger's work, but some idea of its scope and variety may perhaps be gathered from the papers reprinted in the volume *Karl Menger: Selected Papers in Logic and Foundations, Didactics, Economics*, no. 10 of the present collection, published in 1979. A complete bibliography of Menger's work through 1979 may be found in that volume. In a review of the volume in the June 1980 *Mathematical Reviews*, the noted philosopher of science Mario Bunge wrote: "The author's publications span a half-century period and they cover an amazing variety of fields, from logic to set theory to geometry to analysis to the didactics of mathematics to economics to Gulliver's interest in mathematics . . . This selection should be of interest to foundational workers, mathematics teachers, philosophers, and all those who suffer from nostalgia for the times when mathematicians, like philosophers, were interested in all conceptual problems."

Menger's last published paper. "On social groups and relations", appeared in the journal *Mathematical Social Sciences*, vol. 6, 1983, pp. 13-26. It extended work he had begun in his 1934 book on ethics and continued in a 1938 paper in the *American Journal of Sociology*.

On June 2, 1975, in a ceremony at the Illinois Institute of Technology, the Austrian Consul in Chicago presented the Austrian Cross of Honor for Science and Art First Class to Karl Menger, by then Professor Emeritus. The award pleased Menger immensely, for the same honor had been awarded to his father many years before. I.I.T. itself honored him with a Doctorate of Humane Letters and Sciences in December, 1983.

He had a great love of music. Although he found Wagner's political and social views abhorrent, he greatly admired Wagner as a composer, and also as a poet, for reviving in the *Ring* the ancient alliterative verse forms. His favorite Wagnerian opera was *Die Meistersinger*. But his appreciation extended well beyond 19th century opera and encompassed composers from Palestrina to Webern. His enthusiasm, when something pleased him, could be unbounded. Attending a performance of *Arabella* in Chicago in October, 1984, he was smitten by the singing and acting of Kiri Te Kanawa, and exclaimed: "She has the true Viennese style! And she's from the antipodes!"

Over the years, he built up a notable collection of decorative tiles from all over the world. He liked modern architecture. He disliked wine, but enjoyed sweet liqueurs. Though not a vegetarian as has sometimes been said, he ate meat sparingly, particularly in his last years. But he was always glad to sample cuisines, from Cuban to Ethiopian, that were new to him. He liked baked apples.

Until a heart condition curtailed his physical activities he loved to walk long distances. At least once, he admitted to being superstitious. For a long time after the death of Wald in a plane crash, he refused to fly; he relented only after it became practically impossible to travel otherwise.

He died in his sleep on October 5, 1985 at the home of his daughter Rosemary and son-in-law Richard Gilmore in Highland Park, Illinois.

KARL MENGER'S CONTRIBUTION TO THE SOCIAL SCIENCES

In his contribution to the social sciences K. Menger strictly adhered to the canons of methodological individualism characteristic of the Austrian School of Economics. This was, however, married to an unprecedented stress on the need for rigorous logico-deductive mathematical reasoning. By the early thirties, when Menger was most active in this area, emphasis on formalization could hardly be said to be a novelty. The Marginalist School of Economics, represented in various countries, had indicated the use of mathematics in social analysis as a necessary step if it was to become a science. However, the model of mathematization employed was borrowed from classical physics and, more particularly, from mechanics: hence, its mathematics coincided with differential calculus and algebra. (This identification sparked off an impassioned debate with the founders of the Austrian School of Economics and generated a number of misunderstandings.) In his writings and by his personal influence, K. Menger advocated and actively promoted a major shift of formal methods in the direction of mathematical logic and involving the replacement of classical functional analysis by modern topology and combinatorics. In the social sciences, this approach could win a favorable reception only in an intellectual milieu like the one created by K. Menger's own father, where social sciences were founded on two principles — deduction and individual introspection

— which made them essentially different from the hard, experimental sciences. Obviously, so sharp a distinction was no longer felt appropriate in the thirties by members of the Schlick Circle, like the young Menger. Menger's theory of ethics and the related theory of social groups aim essentially at demonstrating the usefulness of logico-deductive, and hence value-free, reasoning and at illustrating its internal rules[1]. His acceptance of methodological individualism requires that ethics and responsibility can only be founded upon the individual's own decisions. His formalized treatment of the creation of social groups and of the emergence of social institutions pioneered later developments in the game-theoretic approach to political economy. Furthermore, in his discussion of the so-called St. Petersburg paradox, Menger must be credited with having pioneered the introduction of subjective probability into the realm of value theory. This technical innovation in utility theory was destined to undergo great developments later on in decision theory and in the theory of games.[2] Likewise, his contribution to the clarification of the so-called law of diminishing returns anticipated modern analyses of the properties of production functions.[3] As Menger himself emphasized, the aim of this article was to exhibit what is demanded by a rigorous logical approach in proving propositions, and thus to indicate the requisites for a meta-theory, rather than to state properties of empirical laws.

On the whole, however, Menger's direct contribution to economic theory was rather limited, and certainly was perceived only much later. In contrast, his role in the development of the modern approach to mathematical modeling was remarkable and, unfortunately, has generally been underestimated. This essentially methodological contribution betrays the unmistakable touch of a first-class mathematician deeply involved in the debate on the foundations of mathematics raging around the beginning of the 30's. Menger's profound knowledge of the three positions represented (Intuitionist, Logicist and Formalist) issued in the instrumentalist point of view denominated "Logical Tolerance", according to which mathematical methods had to be judged only on the basis of their results. It was basically a version of this methodological position, though tainted by a strong Formalistic emphasis, that effectively passed into mathematical economics. The modern foundations of this last were to be laid only after the Second World War by such people as Debreu, Arrow and McKenzie, whose outlook was deeply influenced by post-Gödelian mathematical approaches, in particular by those of the Bourbaki school. However these foundations began to take shape in Menger's Mathematisches Kolloquium at the hands of some of Menger's own friends and students such as Schlesinger, Gödel and, more importantly, A. Wald, largely under Menger's inspiration and at his initiative. As is well known to economists, the Kolloquium was the setting in which there took place the definitive debate on the model of General Economic Equilibrium associated with the Swedish economist G. Cassel.[4] The re-examination of that model had been initiated elsewhere and had some important ramifications involving another first-class mathematician, J. von Neumann. However, in the direction which the debate took in Vienna, for the first time in the history of economics, the issue of rigorously proving the existence of economically acceptable solutions to a set of mathematical equations was proposed as the test of internal consistency of its basic hypothesis: this was treated as an "existence problem" in the technical sense of Hilbert's Mathematical Formalism. The discussion was effectively used to indicate that mathematical formalization in an applied science should be a deductive exercise, subject to the laws of logic and formally separate from the issue of empirical relevance.

Menger reconstructs his role in this revolutionary phase in the re-thinking of economics and gives his views about the early Austrian School in a fine article which appeared in 1973 in a book for the centennial of the publication of his father's *Grundsätze*.[5] The circumstance of being the son of one of the founders of the Austrian School of Economics must surely have played an important part in determining Karl Menger's lifelong interest in the social sciences. However, a keen interest in the relationship between social sciences, hard sciences, mathematics and logic did play a central role in the cultural life of Vienna at the end of the 20's and at the beginning of 30's and regularly had its reverberations in the Schlick Circle. K. Menger played an active part in that general movement within mathematics which proposed its methodology as the canonical model for all other sciences. Probably, Menger thought of carving out for himself a role in the social sciences, and in economics in particular, as the initiator of such a movement towards "exact logical thinking". His contribution lies in having pioneered a methodological direction that has since been generally accepted: a methodology based upon the distinction that he theoretically posited between metaeconomics and economics, and thus between theory and applied economics. A just appraisal of this contribution is not at all easy: economics (and the social sciences generally) seem to have recently taken a sudden turn away from the strict Formalist (and Bourbakist) approach. They now experiment with the new mathematical methods suggested by modern computing facilities. Once again this development echoes similar developments in the contemporary mathematics of complex systems: economics is in the middle of this process which the intellectually adventurous Menger would surely have welcomed.

L. Punzo

Notes

[1] English translation as: *Morality, Decision and Social Organization: toward a Logic of Ethics*, Vienna Circle Collection 6, Dordrecht and Boston, D. Reidel Publishing Co., 1974, originally published in 1933.

[2] "Das Unsicherheitsmoment in der Wertlehre", *Zeitschrift für Nationalökonomie*, 1934.

[3] "Bemerkungen zu den Ertragsgesetzen", Ibid, 1936, English translation in Menger 1.

[4] reported in *Ergebnisse*, various years.

[5] "Austrian Marginalism and Mathematical Economics": J. R. Hicks and W. Weber *Carl Menger and the Austrian school of economics*, Oxford, Clarendon Press, 1973.

EDITORS' NOTE ON THE TEXT

Karl Menger died before putting the final touches to this memoir and had been for some time rather pessimistic in his observations on its state of preparation. It was therefore a matter of agreeable surprise to his friends and heirs both that it was so near completion and that he had clearly indicated a wish that it be published. The collection, transcription, and preparation of a fair copy of the material was the work of Dr. Louise Golland. The material was put into her hands by Professor Abe Sklar, who in turn had been assisted in assembling it by Mrs Rosemary Gilmore, the older of Menger's daughters. All the normal tasks of final editing were then discharged by Dr. Golland in collaboration with Professor Brian McGuinness, who is also the Volume Editor within the Vienna Circle Collection, for which Menger always intended the memoir. In the nature of things it is impossible to identify now and thank all those colleagues, pupils, and members of his own family who (as is known generically) helped Menger in the composition of the work.

I. THE HISTORICAL BACKGROUND

1.

In the fall of 1927, after a stay of two and a half years in Amsterdam, I accepted the chair of geometry at the University of Vienna and returned to my home town. Formerly the capital of a large, multi-lingual empire, Vienna emerged from World War I as the head of a small and poor country that had to go through terrible ordeals. But the famine that had begun during the last war years subsided in 1921; tuberculosis, which had been so rampant during that dreadful period that some physicians called it the Viennese disease, receded again; the run-away inflation was stopped in 1923 when 13,000 old crowns were converted into 1 new shilling; and from that time on, Vienna recovered with amazing speed.

During the years of recovery, the conservative-to-reactionary Christian-Social (*Christlichsozial*) party under Monsignor Seipel governed the Austrian nation, while the city of Vienna with approximately one-third of the country's population was run by the left-socialist, but definitely democratic, party of the Social Democrats (*Sozialdemokraten*).

Though they were unable to carry out the nation-wide socialization of industry planned by their intellectual leader Otto Bauer, the Social Democrats profited from the *de facto* socialization of the real estate in the city resulting from the inflation. Practically all Viennese lived in rented apartments; and by law, house owners could not charge more than the shilling equivalent of the pre-war rent in crowns and thus actually 1/13,000 of the pre-war rent. One consequence was a certain luxury in living space despite the permanent pauperization of large sections of the middle class; another was the unwillingness and inability of the owners to keep up their houses, which were therefore utterly neglected, while no one built new houses. (I miraculously escaped an accident when almost half a ton of stucco fell from the upper part of a house onto the street two steps in front of me.) So, despite an almost 20% drop in the population, mainly due to the emigration of Viennese Czechs to Prague, the pre-war shortage of housing was further aggravated.

Under its extremely efficient minister of finances, Hugo Breitner, the city levied stiff taxes, the revenues of which were used for the building of large housing units, mainly for workers, among them some forbidding-looking, castle-like structures. The city's outstanding department of health and welfare was headed by Julius Tandler, a professor of anatomy at the University of Vienna. Elementary schools were thoroughly reformed, and the centers of advanced adult education, including the *Volkshochschule* (People's University), were exemplary. Sports organizations and guided tours sponsored by the Social Democrats promoted the workers' enjoyment of nature and art.

So, in the mid 1920's, at long last the traditional optimism of the Viennese reappeared. They had overcome the war and the collapse of the empire, famine, disease, and inflation. The despondency of the post-war years gave way to hopes for the future.

2.

During the second half of the nineteenth century, an English-type Liberal party had been very powerful in Austria, but at the end of the century it declined under the onslaught of Socialism, Catholicism, and Nationalism; and after World War I for all practical purposes, the Liberal Party disappeared politically. A small Communist Party was completely overshadowed by the left-oriented Social Democrats and never played a significant role in Austria.

There existed, however, besides the Christian-Socials and the Social Democrats a small, but highly influential party: the German Nationalists (*Deutschnationale*), the direct predecessors of the Nazis, whose main aims were union with Germany (the '*Anschluss*') and antisemitic and antisocialist measures. This party had strong adherents among the minor officers of the bureaucracy and in the non-Catholic minority of the peasantry. Most Catholics, however, were against joining predominantly Protestant Germany.

Among the strongest proponents of the *Anschluss* were people who had come from, or traced their origin to, Bohemia and Moravia, the provinces of the monarchy which became the main parts of the Czechoslovakian Republic after World War I, in particular, those who came from the Sudeten, the mountains that separate Bohemia from Germany on three sides. This group included even some of the leaders, both Jewish and non-Jewish, of the Social Democrats, much as they were, of course, opposed to the other points of the German Nationalist program.

German nationalism had already begun to flourish in Vienna at the end of the nineteenth century. In the time of the Austrian monarchy it was also directed against the Czechs, Southern Slavs, and Italians. Before World War I, there must have been close to a quarter

of a million Czechs in Vienna. Yet there was only one Czech
language school, named after the great Czech educator Komensky
(Comenius). German nationalists constantly interfered with the opera- .
tion of that school. These and similar actions of course prompted
reactions.

Frequently, rather paradoxical situations arose. At one time, the
leader of the anti-Czech German nationalists had the Czech name
Wotawa, while his anti-German Czech opponent had the German
name Rigler. Some purely Slavonic names were Germanized, as
Blasek to Blaschke, others such as Srbik were not changed, although some bearers of such names were militant German nationalists.

Among pre-war philosophers one of the most nationalistic was
the great pre-Wittgenstein critic of language, Fritz Mauthner, who was
born and reared as a Jew in Bohemia, studied in Prague and then
emigrated to Berlin since Vienna was not pro-German enough for his
taste.

3.

In the mid 1920's the points of greatest concentration of German
nationalism in Austria were the universities and polytechnic institutes.

The University of Vienna is divided into several schools (called
Fakultäten): the School of Philosophy including, besides philosophy
proper, all sciences and humanities; the School of Jurisprudence
and Economics; the School of Medicine; and Schools of Theology,
both Catholic and Protestant. These schools take turns in supplying
the president (called *Rektor*) of the university for a one-year term. In
the 1920's, the faculty of the university consisted of German National-

ists, a Christian-Social minority, and a handful of English-style liberals and Social Democrats.

The School of Medicine was less politically oriented than other schools. Yet even there, as the whole world knows by now, Sigmund Freud met with strong opposition, not only because some serious scientists, including Julius von Wagner-Jauregg, objected to unscientific elements of his theory and others were unreceptive to the originality of his ideas, but also because of antisemitism. In the School of Economics, in the 1920's Othmar Spann, an ultranationalistic fanatic, preached a romantic holism (i.e. a philosophy of entirety, of the State, and the Nation) while branding all studies of individual economic wants and acts as 'un-German.' In the School of Philosophy in the mid-twenties, the nationalistic majority was still inclined to moderation; but ominous portents worried the few openminded liberals: these included the botanist Richard Wettstein among the older members of the faculty, and the philosopher Moritz Schlick and the physicist Hans Thirring among the post-war additions. One prominent faculty member in the School of Philosophy, the mathematician Hans Hahn, was an outspoken socialist.

The mentality of the student body reflected that of the faculty, except that the proportion of Social Democrats was of course larger among the students than among the professors. Members of German Nationalist student organizations, their faces adorned with duelling scars, beat up Social Democrat and Jewish students and harassed, where they saw a chance, even members of Catholic fraternities. These incidents led repeatedly to the closing of the university for short periods. But in the spring of 1927, an energetic man was elected rector. He promised to restore order and in fact, by strong measures against the rioters, succeeded in keeping the University of Vienna open during the academic year of his office. He was the

Catholic theologian Theodor Innitzer, a priest who became cardinal of Vienna a few years later.

4.

One dark cloud cast a shadow on the fairly bright landscape. The relations between the two major parties were constantly under bad strain — much worse than is normal in two-party systems — and the daily papers were filled with attacks and recriminations. But the first really serious incident was the result of a failure of justice.

In the middle 1920's, a strange and unprecedented phenomenon became more and more conspicuous in the court rooms of Vienna. The juries returned verdicts of 'not guilty', however blatant the crimes. If a man had committed some misdemeanor, a good defense lawyer would exaggerate the transgression to the point where a jury trial was required, because then his client was sure to go free. Criminal justice was paralysed.

In the spring of 1927, government forces, either military or police, allegedly responding to provocation, shot and killed some workers during a Social-Democratic celebration in a village east of Vienna. Party circles were aroused to the deepest indignation. But when the case came before a jury in July 1927, the defendants were found 'not guilty' and set at liberty. This caused a wild outburst in the Viennese population. Completely disregarding party discipline, which normally was very strict, unorganized groups of thousands of shouting workers rushed into the center of the city and set the Palace of Justice afire. The police quelled the riots within a couple of days. Each side accused the other of the utmost brutality, and in fact, many people were killed.

Most cataclysmic developments are preceded by severe tremors after which calm is restored for some time. The turmoil in July 1927 was such a tremor in the history of the Austrian republic.

II. THE CULTURAL BACKGROUND

1.

Austrian liberalism had produced remarkable cultural results during the second half of the 19th century and up to World War I. The Vienna medical school, second to none, was world famous, even though Freud had to develop his ideas outside of it. Outstanding scientists and historians taught at the University of Vienna. The geologist Eduard Suess combined theoretical discoveries with great practical achievements. The physicists Ernst Mach and Ludwig Boltzmann and the Austrian economic school of marginalists were internationally acclaimed.

At the end of the century, Vienna brought forth writers of typically Austrian cast, some of them — Arthur Schnitzler, Hugo von Hofmannsthal, Hermann Bahr — of international stature. Gustav Mahler made Vienna his home. Young painters broke out of the academic rut in a movement called the *Secession*: among them was Gustav Klimt, too little known outside of Austria. Artistic craft of a new type was produced in the *Wiener Werkstätte* (Vienna Workshop), which still flourished in the 1920's; and Adolf Loos introduced functionalism into Viennese architecture.

2.

The Vienna of 1927 had again become a highly interesting and intellectually lively city. When I arrived in September the town had outwardly calmed down and recovered from the turmoil of July.

There were theatrical and musical performances of the highest quality. The State Opera (with Richard Strauss a frequent conductor) was unsurpassed anywhere. Foreigners who observed Viennese life at that time were amazed at the number of private as well as public recitals of piano and chamber music; and so it became widely known that in Vienna a larger proportion of the population than in most other cities enjoyed classical music. But even more remarkable was the atmosphere at those concerts; at the yearly series of Beethoven quartets it was one of almost religious devotion. Of course, all that was in the Viennese tradition since Haydn and Mozart, Beethoven and Schubert. In the 1920's, the city was at least temporarily, the home of Arnold Schönberg and his disciples, and at the other end of the musical spectrum, of a third generation of composers of popular operettas.

Less widely known is another feature of the cultural life of Vienna, which was a heritage of the politically defunct liberalism and was cultivated by the Social Democrats — the unusually large proportion of professional and business people interested in intellectual achievement. Many members of the legal, financial, and business world; publishers and journalists, physicians and engineers took intense interest in the work of scholars of various kinds. They created an intellectual atmosphere which, I have always felt, few cities enjoyed — a feeling that Percy W. Bridgman at Harvard in a way confirmed.[1]

The numerous members of that intellectual community were extremely hospitable to foreigners — as a rule, more so than to one another — although they occasionally considered guests even from larger cities a bit provincial. The unbalanced post-war combination of exaggerated feelings of economic-political inferiority and cultural superiority had given way to a more stable and dignified appreciation of heritage and potentialities. So all foreign guests seemed to enjoy their stay in the city immensely. Most of the Viennese themselves, given to habitual fault finding had more reservations when judging their home town. But in this I personally was a rather untypical Viennese, and deeply and openly loved the Vienna of 1927.

3.

Supreme prestige attached to the University and its faculty at that time. The Medical School, still first-rate, made Vienna a Mecca for Central and Eastern European patients and physicians. Wagner-Jauregg had just received the Nobel prize for his treatment of paralytic dementia using the organism that causes malaria; Clemens von Pirquet had begun the study of allergies; Karl Friedrich Wenckebach was developing drugs against heart disease.

In the School of Philosophy, outstanding botantists and geologists, historians and philologists continued old trends, while the psychologist, Karl Bühler cooperated with the reform of the schools and teacher training.

In the Law School, Hans Kelsen (who drafted the constitution of the Austrian republic), and one of his students were doing pioneering work on the 'pure' theory of law, studying the structure of the legal system — not so much the contents of laws as the steps, beginning with a basic norm, by which they become laws.

The situation in the field of economics was less favorable. Josef Schumpeter unfortunately had left Austria, after a not so very successful career as minister of finance; and Hans Mayer, in the chair of Carl Menger, the founder of the Austrian school of marginalists, spent most of his time and energy on more or less futile attempts to counteract the holists. The American economist Henry Schultz, who had visited Vienna for a few days in that period, told me that the holist group had asked him only one question: Did he attack economic problems from the point of view of the whole or of the parts? To which, in an epitome of American pragmatism, Schultz had replied, "I attack a problem from a point of view that promises to yield a solution."

On the serious side, Ludwig von Mises gave stimulating lectures without, however, clearly separating the ideas of economic theory (which he presented with an idiosyncratic opposition to the use of even simple mathematics) from his ideal of complete *laissez-faire*. But, as I noticed on several occasions, Mises and his group, who had only little influence on the economic policies of the Austrian government, had deep insight into the theories of money, banking, commerce and international trade, an insight which far transcended Austrian conditions. In particular, I shall never forget the Cassandra-like prognosis early in 1928 of a lady close to the group, who said, when the conversation at a party turned to the fabulous prosperity in the United States and the climbing stock prices in New York: "If the present policies — credit, inflation and all — continue, then the whole story may end in a crash such as the world has never seen before."

4.

Vienna in the 1920's harbored several intellectual and near-intellectual movements in which inordinate numbers of people took an active part.

There was, for example, the group of admirers of Josef Popper-Lynkeus, a surviving friend of Mach, who advocated universal conscription for peaceful instead of for military purposes ('Allgemeine Nahrpflicht statt allgemeiner Wehrpflicht').

A large group worshipped the writer and lecturer Karl Kraus, a very good, if rather vain, satirist, who relentlessly hunted for and exposed empty phrases in newspapers and political speeches as well as symptoms of hollowness and insincerity in works of literature. "Speaking and thinking are one. The speech of inferior journalists (Schmocks) is as corrupt as their thought; and they write in the same ways as they speak," wrote Kraus in his book *Die Sprache* (Language). But without broaching philosophic problems the book criticizes a sloppiness in journalistic style and grammar that readers of printed English can hardly imagine. (Some of his notes are devoted to sentences analogous to "This is bigger as that," "This is as small than that," and "I am looking toward backward.")

Some Viennese intellectuals of the 1920's, particularly women and Jews, were greatly influenced by the writings of Otto Weininger. In his main work, *Sex and Character*, published in the first decade of this century, Weininger asserted the existence of both male and female elements in every human being. He proclaimed the inferiority of woman and Jews in the most passionate terms (though stating somewhere in the middle of the book that he himself was Jewish). He committed suicide soon after the publication of that book, which though it professed certain unpleasant theses, was written with sincerity and even in a definitely ethical spirit. For example, Weininger (and Kraus after him) "reminded an epoch given to judge life as well as art by one-sided aesthetical canons that the morality of an artist is vital to his work."[2] According to Brian McGuinness, Ludwig Wittgenstein admired both Weininger and Kraus. He is reported to have read Kraus' periodical *Die Fackel* (The Torch) regularly.[3]

There was also a large group of followers of Count Coudenhove-Calergi, who studied philosophy at the University of Vienna under Adolf Stöhr and then wrote popular books on ethics and sociology in a characteristic staccato style. In the mid 20's, Coudenhove began propagating the formation of *Paneuropa* — a political union of all the continental European nations west of Russia that would be in a position to deal with the other world powers on equal terms. Just as Demosthenes eloquently but unsuccessfully warned the Greeks, who were politically fragmented into little city states, that they would fall prey to Philip of Macedon unless they united; so Coudenhove warned the Europeans that they would lose their freedom to Russia unless they united in Paneuropa. In 1929, I met Coudenhove on a vacation and, after a few words, he inquired whether I was a member of the Paneuropa Society. When I told him that I was not, he asked in a serious, almost severe tone, "Why not?" I explained that, while sympathizing with some of his ideals, I saw insurmountable obstacles to their realization, especially the obstruction that England would offer in continuance of her centuries-old policy of dividing the continent. He shrugged his shoulders. It took a second world war, with the break-up of the British empire in its aftermath, and the technological development of the first and second third of this century to make at least a large part of Coudenhove's prophetic proposal a reality.

5.

Remarks about the intellectual Vienna of the 1920's would be incomplete without a mention of still another group — one committed to the study of parapsychological phenomena. It may seem strange that this topic could play a role, if only as an apple of discord, even in groups with a strictly scientific orientation; but a partial explanation lies in the circumstances under which that committee originated.

In the first post-war years numerous mediums had appeared in Vienna and they were viewed by the intelligentsia with the utmost skepticism. Finally, one day in the early 1920's, the newspapers claimed that two physicists — professors at the University of Vienna, one tall and one short — had exposed the entire spiritualistic swindle. What had happened, the paper elaborated, was that the short physicist had invited many people to his house for a séance. The guests, sitting in a circle and holding hands in a dark room, had clearly observed the phenomenon of levitation: more specifically, a figure in white rising from the floor several feet into the air. But at that moment the host unexpectedly turned on the lights and everyone could see that the apparition was none other than the tall physicist who, in the dark, had managed to cover himself with a bed sheet and climb on a chair. In the midst of general laughter the two physicists claimed that they had produced a levitation and exposed the mediums.

It goes without saying that the parapsychological groups were outraged; and for once, in a reversal of the ordinary situation, the mediums called all scientists swindlers. But there was also great indignation in the intellectual community; and a group including Wagner-Jauregg, Schlick, Hahn, Thirring, and many others (most of them scientists) formed a committee for the serious investigation of mediums. Very soon, however, members began to drop out: first, Wagner-Jauregg; soon after him, Schlick; so that by 1927 only two of the scientists were left, my friends and former teachers Hahn and Thirring. They were not fully convinced that any of the phenomena produced by the mediums were genuine; but they were even less sure that all of them were not. They believed, rather, that some of the parapsychological claims might well be justified; and that certainly the matter warranted further serious investigation.

Even though Hahn tried to get me actively interested in the work of the committee I stayed away except for one short contact. This occurred when they brought in a young woman from Graz (the capital of Styria), who was reported to have evoked astounding raps in distant objects. The committee planned séances with her. At the last moment Hahn was prevented from taking part in one of these and asked me, as a personal favor, to attend and report to him.

I found the hotel room, which had been rented for the girl and her aunt, well lit. But to my amazement — at first I thought I had entered the wrong room — the guests were rhythmically chanting the words of the children's song *O Tannenbaum*. This, someone told me in a whisper, stimulated the phemomena, which occurred to the rhythm of this song. The sullen-faced girl was half sitting, half lying on a large wooden bed, near the front end. I took a chair quite close to her. After a while she said dreamily, "It is coming"; and after another short time one could hear raps (first single, then in the rhythm of the children's song) emerging from the wooden board at the head of the bed while the girl stayed quietly at its foot. It sounded as though a very heavy caterpillar was knocking its head against the walls of a passage that it had eaten inside the wood. Then the raps started wandering in the board towards where I was sitting. However, just when I began to be rather impressed, the phenomena deteriorated and after a while completely stopped. Some of the guests asked questions to be answered either by Yes (one tap) or No (two taps); and by this cumbersome procedure, in which the girl seemed to be rather uninterested, it was ascertained that some skeptic in the audience was the cause of obstruction. Then questions followed, "Is it Mr. A?"; "No" "Is it Mr. B?", "No"; and so on. But when my name came up the answer was one knock. Of course I immediately got up and left the room even though the other participants wanted me to stay.

On the way home I reflected on how totally unqualified I was to deal with phenomena such as those produced by this medium. On the few occasions when I had seen magicians in variety shows I abandoned myself to a passive enjoyment of what I saw or heard without ever trying to explain their tricks. I certainly could not explain the knocks that I had heard, whatever may have caused them. I decided never to visit a séance again.

6.

A remarkable feature of Vienna in those years was the existence of numerous *Kreise* (circles, in the sense of discussion groups), some with, some without, direct ties with the academic world.

There was of course a large number of socialistic discussion groups (one conducted by Max Adler was totally Kant-oriented), and psychoanalytic circles of various factions, which were attracting visitors in ever-increasing numbers, especially from the English-speaking world, and groups devoted to debates on educational reforms.

Discussion circles connected with the university included, among others, one conducted by Kelsen and one by Mises. To the Mises

Circle belonged Gottfried von Haberler, Friedrich A. von Hayek, and Fritz Machlup, who were later to continue the tradition of Austrian marginalism in the English-speaking world.

Particularly numerous were philosophical circles: one, directed by Professor Heinrich Gomperz, was historically oriented; others dealt with Immanuel Kant, with Søren Kierkegaard and Leo Tolstoy (the latter's influence was enormous); and some circles discussed the philosophy of religion; and others phenomenology.

I knew about these circles since most of them included one or another of my friends; but I personally did not belong to any circle, except one that was non-scientific. Until one day in the fall of 1927 Hahn said to me: "I hope you can join our philosophical circle round Schlick. We meet informally about every other Thursday evening in Schlick's Institute. Rudolf Carnap, Otto Neurath, and some younger people attend regularly."

"Schlick has not seen me since my doctoral examination in 1924," I answered, "I doubt that he remembers me."

"I have already spoken to him," Hahn said. "He remembers you well and is pleased with the idea that you might join us."

I thanked Hahn and accepted with anticipation.

Of the many intellectual circles that flourished in Vienna in the 1920's, it was the one that its members called the *Schlick-Kreis* which was to become known throughout the world as *Der Wiener Kreis* or *The Vienna Circle*.

Notes

[1][Probably Percy W. Bridgman, *Review of Scientific Instruments*, v. 4, 1933 L. G.]

[2]Cf. P. Engelmann, *Letters from Ludwig Wittgenstein with a Memoir*, Oxford, 1967, p. 125 sq.

[3]Ibid, p. 138.

III. THE PHILOSOPHICAL ATMOSPHERE IN VIENNA

1.

The philosophical atmosphere of a city or a country is a rather elusive sociological phenomenon. The opinions professed in universities are by no means all-decisive. Historical factors contribute to the atmosphere; for works from the past are discussed more widely and with greater understanding in the place of their origin, where disciples and disciples of disciples of the creators may still be active. Philosophical societies play a role. Informal conversations between intellectuals when they talk about philosophy (which they did abundantly in the Central Europe of the 1920's) contribute to the climate; and what students say is of some interest. The diverse nature of these elements makes it hard to avoid some desultoriness in the description of the atmosphere. Yet by following the development of all those factors in Vienna one may understand why the *Schlick-Kreis* blossomed just there.

The dominant historical element is of course the fact that Austrians never contributed to the German type of metaphysics that culminated in Fichte, Schelling, and Hegel. The great thinkers born in the Austrian empire, Bolzano and Mach, used to philosophize along scientific lines.

2.

The logician, mathematician, and methodologist Bernhard Bolzano (1781-1848) was the most important forerunner of Georg Cantor, the founder of set theory, and of Karl Weierstrass, the creator of a rigorous theory of real functions. While the metaphysicians were indulging in speculations about 'the absolute', about infinity, and about continuity, Bolzano soberly studied infinite classes, did away with Euclid's axiom that the whole is necessarily greater than the part, and investigated one-to-one correspondence between sets.[1] He had a sound definition of continuous functions and described them in theorems supplied with rigorous proofs. He even anticipated Weierstrass in defining a continuous function that is nowhere differentiable, or, geometrically speaking, in the construction of a continuous curve without any tangent — something that even Weierstrass' contemporaries in the 1870's considered as totally contradictory to geometrical intuition. Figures III.1.a, III.1.b, III.1.c show three successive approximations to such a curve.

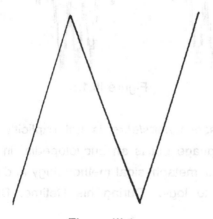

Figure III.1.a

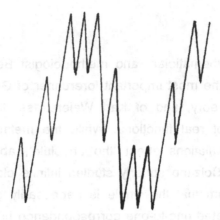

Figure III.1.b

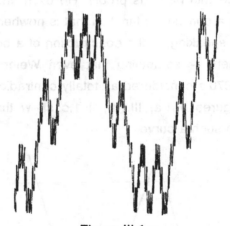

Figure III.1.c

Bolzano's *Wissenschaftslehre* is not explicitly critical of meta-physics or of language but is an encyclopedia (in four volumes) of pragmatic, totally unmetaphysical methodology and contains interest-ing contributions to logic. During his lifetime, Bolzano remained practically unknown. But at the beginning of this century, his works began to become known and greatly appreciated in Austria.

3.

The physicist Ernst Mach (1838-1916) developed a more modern and much more radical positivism than that of Auguste Comte in France or Herbert Spencer in England. He analysed his sensations into elements, and showed how physical bodies, other humans, and his own self could be constituted by certain comparatively stable complexes of such elements. The 'Introductory Remarks' to his book *Contributions to the Analysis of Sensation* (1885) have the subtitle 'Antimetaphysical.' In these remarks, Mach attacked Kants's "monstrous notion of a *thing in itself*, unknowable and different from its *phenomenal existence*." He called some traditional philosophical questions *pseudo-problems.* He furthermore proposed a combined and unified study of physical, physiological, and psychological material. (I remember that years ago I read with amusement an essay in which Freud seemed to claim implicitly a monopoly of the term *analysis*, while speaking rather disparagingly of Mach's book.)

Mach interpreted science and mathematics from a point of view of intellectual economy (Denkökonomie) considering them as labor-saving methods. With regard to causality, Mach agreed with David Hume's skepticism. Other ideas of his, (especially the denial of the existence of external things) and even his ways of expressing them are very similar to Berkeley's, as Karl Popper[2] documented in detail a few years ago. The similarity had been clearly seen by Lenin who, however, connected Mach with Bishop Berkeley even beyond scientific points of contact — a fact that had led to a distrust of positivism, and at times to its condemnation, in the Soviet Union.

Mach was against models in physics, as was Pierre Duhem, who called them parasitical elements of explanations; and in this train of thought, Mach strongly opposed atomism. It is true that most of the experimental support of atomism was discovered after his time; yet I have never understood how Mach could ignore the fact that the alternative — the assumption of continuous distribution of matter — is likewise based on unproven and indeed unprovable assumptions. The granular structure of matter, once experimentally ascertained, would remain established forever; whereas the assumption of a continuous distribution would always be threatened by the possiblity that refined experimental techniques might later disprove it.

One of Mach's greatest achievements, which made him a pre-cursor of Albert Einstein, was his early insistence on the relativity of all observed motions, even of accelerated motions including rota-tions. Newton had tried to prove that the latter are absolute (i.e. relative to absolute space) by pointing to the centrifugal forces in rotating bodies. The bulging of the equator of a rotating plastic sphere (usually described as the flattening of the sphere at the poles) is an example. Mach pointed out that such a rotation is relative to the fixed stars of distant but immense combined masses, and made the ingenious suggestion that if an enormously heavy ring or shell were rotated, then a plastic sphere 'at rest' in the center might also display a bulging equator. In the early 1920's, Thirring actually deduced this phenomenon from Einstein's general theory of relativity.

It is interesting that, on the other hand, in Chaper 58 of his *Principles of Mathematics* (1902), Bertrand Russell claimed that "Absolute motion is essential for dynamics." He called Mach's suggestion that one could not infer the rotation of the earth if there were no heavenly bodies "the very essence of empiricism, in a sense in which empiricism is radically opposed to the philosophy advocated in this work." And it is ironical that he concludes that chapter with the remark, "Absolute motion is for us a powerful confirmation of the logic on which our discusssions have been based."

Mach wanted to replace all causal explanations by descriptions involving functions — a view that nowadays is even held by some economists. But it should not be overlooked that even physical sciences abound in descriptions of a merely qualitative nature. When we describe thunder as longitudinal air waves following a large electric spark we have come a long way from the explanation involving the blow of an irate god's hammer; and yet that road has not been paved with equations. A weakness that Mach's extreme empiricism shares with John Stuart Mill's, lies in the attempt to base all of arithmetic on experience.

Most of Mach's work was done in Prague, which he left in 1895, when a chair of the philosophy of the inductive sciences was created for him at the University of Vienna, He retired in 1901 at the age of 61.

Mach, who had many friends and followers in the English-speaking world, exerted a great and lasting influence in Austria.

4.

Near the turn of the century, Fritz Mauthner (1848-1923), until then journalist and bellettristic writer, started a massive critique of language. He had studied in Prague. "Mach's epistemological positivism has been subconscious in my mind," he wrote in his Memoirs: and Gershon Weiler says, "The only teacher who was significant for Mauthner's later career was Mach . . . (Austrian philosophy), largely through the influence of Mach, had shown features which marked it off from the general trend of German philosophy. Mauthner, who chose to be a German over being an Austrian, became thus . . . one of the representatives of Austrian spirit in German philosophy."[3] But he hardly made any impact on philosophical thinking in Germany.

Mauthner published three voluminous philosophical works: *Contributions to a Critique of Language* (1902), *Dictionary of Philosophy* (1910,1923), and *Atheism and its History in the Western World* (1922).

All that philosophy can do, according to Mauthner, is to give critical attention to language. Traditional philosophy he described as *word fetishism*. The purpose of language is to facilitate our activities. All our definitions and most of our statements (in fact, all of them except communications of new observations), Mauthner declared to be 'logical tautologies.' But he had no clear concept of tautology and, averse to logic, he was completely unfamiliar with the developments after George Boole and Schröder. Mathematics he considered to be certain but totally uninformative.

In contrast to Wittgenstein after him, Mauthner wrote in a rather superficial feuilletonistic style. But like his successor, he was convinced that he had definitely solved all philosophical problems. As I was told by Edmund Benedikt, a prominent Viennese lawyer and first cousin of Mauthner, that conviction made him insufferably conceited.

Strangely enough Mauthner had a strong inclination to mysticism and tried to reconcile it with his atheistic views. "I will try to say the unsayable and to express in poor words what I can communicate of nominalistic mysticism, of skeptical mysticism," he wrote. "I can experience it in hours . . . hours of ecstasy . . ." Lying in high grass in a quiet summer day he would feel all differences between the world and himself disappear.[4] This mysticism erases the motives of love, greed, and vanity. A booklet *Godless Mysticism* by Mauthner appeared posthumously.

Mauthner was a great admirer of the 14th century German mystic Eckhard, in particular of his sayings about silence. Silence appeared to Mauthner as the culmination of the critique of language.

5.

During the fourth quarter of the last century, the German philosopher Franz Brentano taught at the University of Vienna founding a widely ramified school that was very influential thoughout the Austrian empire: Alexius Meinong and his pupil Ernst Mally taught in Graz; Alois Höfler, another pupil of Meinong's, in Vienna; Christian Freiherr von Ehrenfels, Anton Marty and his pupil Oskar Kraus, in Prague. Kazimierz Twardowski was one of the first Polish philosophers; Thomas Masaryk became the first president of the Czechoslovakian republic after World War I; Edmund Husserl went to Germany. Höfler edited Bolzano's *Wissenschaftslehre*, a book greatly admired by Husserl.

After his retirement, Brentano started a rudimentary critique of language. He was mainly interested in eliminating nouns that do not designate anything and correspond to fictitious entities. Less praiseworthy, though widely acclaimed by his followers, was his work in ethics, which culminated in what purported to be a definition of the *good* as "that which is worthy of love" and "that which can be loved with a love that is correct." The definition was implemented by maxims such as the rule that, of various possible aims, one should always choose the best. I, for one, have never been able to take such empty verbiage seriously.

Meinong had closer connections with British philosophy than with German metaphysics. He wrote extensively about Hume and was in correspondence with G. E. Moore and Russell. Yet his theory of objects (*Gegenstandstheorie*), in which he speaks of an object for every mental act, is in a certain sense the exact opposite of positivism and critique of language. The realm of 'objects' discussed and classified in that theory includes, for example, the present president of France, the present king of France, golden mountains, mermaids, and square circles.

Meinong's theory of objects, which Russell studied with great interest, was the incentive for the development of the latter's theory of descriptions. According to Russell's theory, the proposition "the present king of France is bald" is not as meaningless as other philosophers have claimed, but simply false since, by virtue of the article 'the', the proposition implies that there is one and only one man who is the present king of France, which is not true.

6.

Square circles, which rank even below mermaids in Meinong's hierarchy of objects, warrant a short digression since they have haunted many and diverse philosophical writers. Under various names (including round squares) they have been discussed by Thomists, existentialists, linguistic philosophers as well as by Bergson, Russell and others.

By a circle and, more specifically, by the circle in the plane Π with the center O and radius of r units, philosophers cannot very well have meant anything but the class C of all points in Π that are at the distance of r units from O. But they probably have been unaware of the following facts: 1) that the shape of that circle essentially de-

pends on the method by which distance are measured in the plane
Π, 2) that at the beginning of this century Hermann Minkowski indicated a definition of distance in a plane Π according to which a
circle is indeed a square, and 3) that, as I pointed out many years
ago, people (including philosophers) actually encounter square circles in their lives, e.g. whenever they use a taxicab in a modern city
built in square blocks. Indeed, for a ride from an intersection O to a
point p that is 4 blocks east and 3 blocks north of O the cab driver
will, according to his motor, charge a fare for 7 blocks. To little avail
will the philosophizing passenger protest that in a right triangle
having sides of 3 and 4 units length the distance between the end
points of the hypotenuse is 5, not 7, units, as the ancient Egyptians
already knew and as Pythogoras proved. *He must pay the fare for 7
blocks.* And where are the other points that he could reach at the
same cost, i.e. the other points a taxicab distance of 7 blocks from
O? They are (See Figure III.2.a) the intersection

    3 blocks east and 4 blocks north,

    2 blocks east and 5 blocks north,

    1 block east and 6 blocks north,

    7 blocks due north,

    1 block west and 6 blocks north,

    2 blocks west and 5 blocks north,

    and so on;

altogether 28 points. They lie on a square (in diamond position to the
streets), whose corners are 7 blocks straight north, east, south, and
west of O.

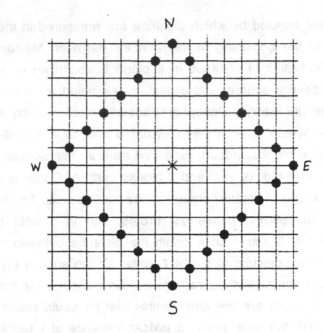

Figure III.2.a

The measuring of the distance in the taxicab geometry thus leads to square circles,[5] which are even of certain practical significance, namely for the passengers' purses.[6]

Clearly these facts are unimportant for philosophers since, for their purposes, they might replace the square circle by another quasi-geometric example (a quadrangular triangle might be of a 'arithmetical' nature). But they must be careful. For while mathematicians cannot associate any meaning with many of the deep assertions of metaphysicians, they have now given a perfectly meaningful interpretation to the paramount example of what philosophers call *meaningless*.

7.

The Philosophical Society of the University of Vienna arranged meetings that were open to the public, and people availed themselves of this opportunity in large numbers. The Society, which during the seond decade of this century was strongly influenced by Höfler (see III, 5), published a *Jahrbuch* that contained interesting contributions. One of the principal topics of discussion was the foundations of mechanics.

One inconspicuous item that I remember reading was a short report about a meeting very early in this century featuring the German geometer Felix Klein as guest speaker. In keeping with his opposition to Cantor's set theory and his negative evaluation of Weierstrass' type of mathematical rigor, Klein made disparaging remarks about the latter's curve without tangents at any point (see III, 2) and similar 'pathological' examples. But the physicist Boltzmann raised his voice in defense of the more modern mathematics and said (referring to the little meteorological towers displaying graphical registrations of temperature and air pressure that stood in all Austrian summer resorts), "I never pass a *'Wetterhäuschen'* without being reminded of Weierstrass' curve without tangents (see Figure III.1.a, III.1.b, III.1.c) by the graphs of temperature and pressure."

This exchange and especially Boltzmann's interesting remark illustrate the kind of discussion that would be unthinkable in a metaphysical atmosphere but was considered most appropriate in a philosophical society in Vienna.

8.

Mach left Prague in 1895 when a chair for the philosophy of the inductive sciences was created for him at the University of Vienna. When he retired in 1901 at the age of 63, he was succeeded by Ludwig Boltzmann until that great physicist's death in 1906. From then until the arrival of Schlick, the fortress was held by the Austrian philosopher Stöhr, who was strongly influenced by Mach and also interested in a certain kind of critique of language. In his *Psychology* he wrote, "If there were no words, there would be no nonsense, at worst, errors. . . Nonsense cannot be thought, it can only be spoken." In 1898, he wrote a book *Algebra of Grammar*, which seems to be worth being examined from a modern point of view. In the autumn of 1920, before he could start his course on philosophy which, on the recommendation of many students I planned to attend, Stöhr fell seriously ill and died.

In 1922, Moritz Schlick took the vacant Mach-Boltzmann-Stöhr chair (see IV,2). The mathematican Hans Hahn, who had come to the university in 1921 (see V,3) and was instrumental in bringing Schlick to Vienna, was greatly interested in logic. He offered a lecture on the algebra of logic in the fall and winter of 1922; and during the academic year 1924-25 he devoted a seminar (jointly with the geometers Kurt Reidemeister, Josef Lense, and Leopold Vietoris) to Russell and Whitehead's *Principia Mathematica*. There was a large attendance, and the participants reported on the various chapters of the book. I attended only the first half of the seminar, since in the spring of 1925 I left Vienna and went to Amsterdam. Together with Schlick's courses on the theory of knowledge and on philosophy of science, Hahn's seminar created the background and the basis for the development of the Vienna Circle.

9.

In my student days in Vienna (1920-24), one could frequently hear discussions concerning the theory of relativity, the foundations of physics and geometry, and related topics — not only in the Philosophical Society and among students of science but also at almost every gathering of members of that numerous scientifically interested intelligentsia made up of professional and business men (see II,2) — and the spirit of Mach prevailed. Because of Boltzmann's great work and his lectures still another point of view became widely known and influential — looking at phenomena in the large as statistical results of events in the small. Good semipopular writings by the physicist Franz Exner of the University of Vienna helped to spread this view. Later Hahn's lectures and seminars brought logic, the foundations of arithmetic, and the work of Gottlob Frege and Russell into the foreground, though mainly among philosophically interested students of mathematics and science.

I had many discussions with my friend and fellow-student of mathematics, Otto Schreier, who was to develop into one of the most outstanding group theorists and whose untimely death in 1929 was a great loss for mathematics. After reading books on traditional philosophy and especially on social science we arrived at the conclusion that most of the numerous controversies in those fields were actually about nothing but definitions; and as mathematicans we felt that all definitions are arbitrary and command acceptance only by convention.

Another topic of discussion, especially when our friend, the phenomenologist Felix Kaufmann, joined us was the existence of meaningless sentences. Kaufmann claimed that besides true propositions such as "snow is white,"and false propositions such as "coal is white,"there were meaningless propositions such as "César Franck's symphony is white." Schreier emphatically denied Kaufmann's view and maintained that the third sentence was simply false: a symphony is *not* white. Later discussions in the Circle seemed to vindicate Kaufmann in this matter although I myself, ever since thinking about this problem, have considered the use of the term 'meaningless' as subject to a good deal of arbitrariness. (I have also always felt that it is easy to lose all common sense in such discussions and therefore found it refreshing when a young female student, who happened to be present at one of them without seeing the point of the controversy, said to Kaufmann, "I don't understand why you insist on that sentence being meaningless. Has anyone *claimed* that Franck's Symphony is white?")

Still other discussions centered on intuition. Kaufmann, an ardent admirer of Husserl, believed that *Wesensschau* (intuitive grasp of essences) was the foundation of various statements and proofs. When I asked for details Kaufmann urged me to study Husserl's *Ideen zu einer reinen Phenomenologie* (Ideas, General Introduction to Phenomenology). But I stumbled right at the beginning, where I read what purported to be a definition of essence: "Essence is that which in the primordial being of a thing is found as its what" (*"Wesen ist das im ureigenen Sein einer Sache als ihr Was vorfindliche"*). Nor was I enlightened when I read that "whatever presents itself in intuition in primordial form is simply to be accepted as it gives itself out to be, though only within the limits in which it presents itself." For I reflected that all continuous curves had presented themselves even to the intuition of the greatest mathematicians before Bolzano and Weierstrass as having tangents (see III,2); and since the pre-Bolzano mathematicans simply accepted the continuous curves "as they gave themselves out to be" they made assertions (e.g. that all such curves have tangents) which were simply false. To be sure, Husserl's words 'in primordial form' especially the phrase "only within the limits in which it presents itself" may be used as an escape to uphold his statement even in view of honest appeals to intuition that are blatantly incorrect. But if so used, those qualifying clauses deprive Husserl's statement of any content.

In my student days, I was greatly confused by Hermann Weyl's praise of Husserl. I had not as yet realized as I did later, after meeting Weyl and hearing him lecture, that that great mathematician had a strong penchant for metaphysics. Schlick after meeting Weyl on a vacation told me that he found him overawed by tradition in the field of philosophy.

In a seminar, the physicist Felix Ehrenhaft read with us Duhem's important book *Aim and Structure of Physical Theories*, which made a deep impression on us. It had been translated into German by the Austrian physicist and politician Fritz Adler twenty years before an English translation appeared. With another friend of mine I read some chapter of Mauthner's three volumes *Kritik der Sprache* (Critique of Language) and we were very much impressed. These and similar conversations and readings undermined my original interest in the classical systems of philosophy.

10.

What we did not know in the early 1920's was that our iconoclastic attitude continued a tradition among Viennese students of philosophy which even then was at least half a century old. This appears from a story by which Brentano illustrated what he considered as the sad state of philosophical thinking in Vienna when he arrived from Germany in 1876. In a meeting of the students of philosophy, the speaker advanced the view that there had been a time for theology, followed by a time for philosophy, until in the 1870's the time had come for students of philosophy to study only exact sciences. One might think Wittgenstein was the speaker (see VII, 2); but he was not yet born. Brentano repeated with deep indignation, "And this in a meeting that not only had been arranged by students of *philosophy*, but to which they had invited *me*." While Brentano did much to promote an unmetaphysical, but also unpositivistic philosophy, a totally positivistic undercurrent remained.

Quite recently, I discovered an even older document attesting to the antimetaphysical spirit of young men in Vienna. Looking through by father's old notes, I found that in 1867, just before developing his interest in economics (which resulted in the publication of his *Principles of Economics* in 1871), he planned to write a book *Critique of Metaphysics and of the so-called Pure Reason from the Empiristic Point of View*. He characterized its objectives in the following seven short paragraphs:

> All so-called ideas *a priori* and knowledge from pure reason must be presented as empirical statements or as errors, i.e. false experiences or empty compilations of words (*leere Wortzusammenstellungen*).
>
> There is no truth of a metaphysical nature and thus lying beyond the limits of experience. Beyond, there are only rational calculi (*Verstandeskalküle*) and fantasies.
>
> There is no metaphysics. There is only a theory of the correct observation and consideration of things that is free of prejudices, be they accidental or created by education.

There are neither *aeternae veritates* as the dogmatists claimed nor forms of perception and thought lying in us by which Kant replaced the *aeternae veritates*.

Kant rejects metaphysics and replaces it by the critique of pure reason. I say, there is no pure reason.

There is no riddle of the world that ought to be solved. There is only incorrect consideration of the world. This objection is directed against the essence of modern philosophy and against the form of empiricism.

Mere materialism has equally pernicious consequences for science as mere idealism. Just as the latter confuses the world, so the former makes it shallow.

Unfortunately, my father never carried out his plan to write a book along these promised lines.

11.

In the intellectual atmosphere of Vienna in the early 1920's, everything seemed to point to one fact: that the stage was set for more systematic discussion on a higher level. Schlick's Circle filled a definite need.

Notes

[1]Bolzano realized that by associating to each of the numbers 1,2,3, . . . twice that number, i.e. the numbers 2,4,6, . . ., respectively, he established a one-to-one correspondence between the whole set of all positive integers and its part consisting of the even integers. Hence that whole cannot be said to be greater than that part. Similarly, if on a scaled line one associates with each point between 0 and 1 the point twice its distance from 1 on the line, that establishes a one-to-one correspondence between the segment from 0 to 1 and the segment from 0 to 2, which contains the

former as a part.

[2]*Brit. J. Phil. Sci.*1954

[3]Gershon Weiler *Mauthner's Critique of Language*, Cambridge: Cambridge University Press, 1970, p. 335.

[4]Mach too, if I remember correctly, mentions that it was on a summer day while he was lying in a meadow that the recognition of the unity of all his sensations came to him — those normally said to come from the outside and those from within. (Central European meadows, studded with hundreds of flowers every summer, are indeed unique, and to lie there in the grass is a most beautiful experience. Nowhere in the United States nor even in other European countries, such as Holland, have I found meadows of the kind that exist in the lower-lying parts of Austria and Switzerland.)

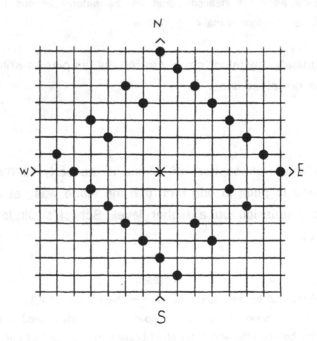

Figure III.2.b

[5]If the streets and avenues are alternately East-bound and West-bound, and North-bound and South-bound (See Fig. III.2.b) the circles are even stranger. If O is e.g. the intersection of a North-bound and an East-bound street, then the 8 points in the northeast quadrant at a distance of 7 blocks are the same as in the case of two-way

streets; but in the northwest and southeast, 8 points of this circle lie on two lines *parallel* to the line joining the 8 points mentioned above. Altogether there are only 24 points on this circle, which is not quite symmetric about O. Whereas its northernmost point is 7 blocks straight north of O, its southernmost point is 6 blocks and 1 block east of O; and there is a similar asymmetry in the east-west direction.

[6] "But the distance travelled along broken paths," a philosopher might object, "Is not what we mean by distance. We measure distances as the crow flies." Measuring distances as the crow flies is indeed a useful convention in most, but not in all situations. A cab moving from O to P between city blocks obviously cannot proceed along the diagonal which those have in mind who say that the distance between O and P is 5 blocks. Nor, for that matter, is there a flight from O to P that would be only 5 blocks long *even for a crow*. For in order to profit from a diagonal shortcut in the air, the crow must first somehow fly above the houses and finally down again. But such a path exceeds the length of 5 blocks; and if the houses are very tall it may even add up to 7 blocks or more.

IV. WHY THE CIRCLE INVITED ME.
THE THEORY OF CURVES AND
DIMENSION THEORY

1.

As Hahn had indicated, Schlick, Carnap and he himself wanted me
to attend the meetings of the Circle because they were interested in
the methodology of my work on curves and dimension. On several
occasions, these matters came up in discussions of the Circle and
they have been referred to in the writings of Carnap and Popper.

2.

In March of 1921, just after Hahn's arrival in Vienna, I read a notice
that he would conduct a two-hour advanced seminar every Wednes-
day during the spring semester, entitled *Neueres über den Kurven-
begriff* (News about the concept of curves). At that time I was at the
end of my first semester at the university and primarily a student of
physics, I was doubtful that I would be able to follow the advanced
mathematical discussions; and I had never seen Hahn before. But
after some deliberation, I decided to try.

The first meeting of the seminar allayed my doubts. Hahn went
right to the heart of the problem. Everyone, he began, has an intuitive

idea of curves: everyone calls a circle and a lemniscate (of the shape of the numeral 8) curves; no one so designates a circular disk or a cube. When speaking of curves one thinks of thread-like objects, wire models, and the like. But anyone who would try to make that idea precise, Hahn said, would encounter great difficulties. In this seminar we would examine attempts by several eminent geometers to define the curve concept, only to find that some of their definitions were too wide, others too narrow and still others altogether unsatisfactory, so that at the end of the seminar we should see that the problem was not yet solved.

I was completely enthralled; and when after that short introduction, Hahn set out to develop the principal tools used in those earlier attempts — the basic concepts of Cantor's point set theory, all totally new to me — I followed with the utmost attention.

If S is any set of points in space, he began,[1] then a point p (which may or may not belong to S) is said to be an *accumulation point* of S if every sphere with center at p, however small, contains infinitely many points of S.[2] A point q is said to be an *interior point* of S, if S includes all points of some sphere with center q. A set R is called *closed* if all its accumulation points are points of R or if R has no accumulation points. A set U is called *open* if all its points are interior points of U or if U contains no points at all (in other words, is empty). Any set consisting of only a finite number of points has no accumulation points and, therefore is closed; so are for example spheres, cubes, pyramids, the surfaces of solids, plane disks, ellipses, straight lines and segments. Solids stripped of all points of their surfaces are open sets.[3] But whereas a door, according to a French saying that occurred to me, must be either closed or open, most sets, Hahn said, are neither. For instance, if V is a cube from whose surface only one single point has been deleted, then V is not closed (because the deleted point is an accumulation point of V, but not a point of V); nor is V open (because each nondeleted point, r, of the surface is a noninterior point of V; each sphere centered at r, however small, contains points not belonging to V). The entire Euclidean space, on the other hand, is a set that is both closed and open — also an impossible state for doors, I noted. If q is a point of the open set U, then U is also said to be a neighborhood of q.

The surface of a cube C, clearly consists of its noninterior points, that is, of those points of C for which each neighborhood, however small, also contains points not belonging to C; the surface may also be said to consist of those accumulation points of the open cube (that is stripped of its surface) which are outside that open set. This latter fact is generalized in the following definition. The *frontier of an open set*, U, consists of those accumulation points of U that are not points of U.

With only about ten minutes of that first two-hour session of the seminar left, Hahn had time to introduce just one more definition. It was providentially, as I later realized, that of *continuum*, an idea which is more intuitive than the notion of closed and open. According to Camille Jordan, a closed set is *not* a continuum if it is the union of two nonempty closed sets that are disjoint (that is, have no point in common) — briefly, if it can be split. The only other closed sets not called continua are sets consisting of a single point. A continuum, then is a closed set containing more than one point which *cannot* be split — which in other words, is not the union of two nonempty close sets that are disjoint.[4] Clearly, a set consisting of only a finite number of points is closed but is not a continuum. Neither is the union of a circle, square plate and a cube that are pairwise disjoint, for this union can be split into two disjoint closed sets; for example, into the

the circle and the union of the plate and the cube. A line segment, a disk and a spherical solid are examples of continua.[5]

3.

I left the seminar room in a daze. Like everyone else, I used the word 'curve' and had an intuitive idea of curves — mental pictures associated with the term. Should I not be able spell out articulately *how* I used the word and to describe clearly what I saw?

After a weekend of complete engrossment in the problem I felt I had arrived at what appeared to be a simple and complete solution. I told the idea to my friend, the highly talented mathematican Schreier, who listened with great interest. The next day he said that he had given much thought to my attempt but could not find any flaw in it. "However," he added, "if an idea as simple as yours could solve the problem, why would several great geometers have given unsatisfactory solutions and why would Hahn say that after discussing all previous attempts we should see that the problem was still unsolved?" I admitted that this thought had also crossed by mind. Schreier continued, "I then looked into the standard work on set theory, *Grundzüge der Mengenlehre* by Felix Hausdorff, the greatest authority on the subject in Germany; and he claims that the sets traditionally called curves are so heterogeneous that they do not fall under any reasonable collective concept." Moreover he showed me that Ludwig Bieberbach in his booklet on differential calculus said: "Anyone trying to define the concept of curve comprehensively and in a way that fully renders our intuitive idea certainly would need a description as long as a tape worm and of Gordian entanglement."[6] "Yet as you will agree," I said in keeping with a general penchant of mine for simplicity, "one should never reason that an idea is too simple to be correct. I shall tell my solution to Hahn."

There was at that time hardly any contact between students and professors at the University of Vienna. So, an hour before the beginning of the second seminar meeting, I had to muster all my courage before knocking at the door of Hahn's office to tell him that I thought I could solve the problem he had formulated the week before.

Curves, like surfaces and solids, I began, are continua of a special kind. If a continuum is a solid, or a surface and p is a point of S, then S has a continuum in common with the frontier of every small neighborhood of p (see Figure IV.1).

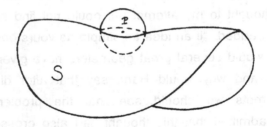

Figure IV.1

If, however, C, is a curve and p is a point of C, then p is contained in neighborhoods as small as one pleases, whose frontiers have *no* continuum in common with C (See Figure IV.2). For example, a curve made of wire pierces the frontiers of some (spherical or cubic or other) neighborhoods as small as one pleases of each of its points only in dispersed points.[7] On the other hand, a paper surface or a wooden solid cuts the frontiers of each sufficiently small neighborhood of each of its points along an entire continuum.

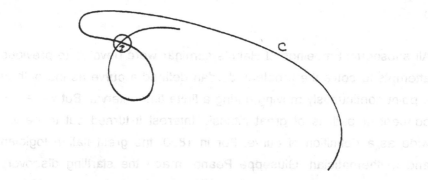

Figure IV.2

Hahn, who hardly looked up from the book he was reading when I entered, became more and more attentive as I went on. At the end, after some thought, he said that this would be indeed a workable definition, and asked me where I had learned so much about point sets and topology. I replied that I was a physics student at the end of my first semester and had not heard about topology; but that my definition used only concepts defined the week before in the first meeting of his seminar, which was all that I had ever heard about point sets.[8] Hahn had not realized that two hours of his excellent presentation of basic concepts were sufficient to make them operative even in the mind of someone totally unfamiliar with the field. He nodded rather encouragingly and I left.

4.

All subsequent meeings of Hahn's seminar were devoted to previous attempts to solve the problem. Jordan defined a curve as the path of a point continuously moving during a finite time interval. But while this concept of path is of great intrinsic interest it turned out to be too wide as a definition of curve. For in 1890, the great Italian logician and mathematician, Giuseppe Peano, made the startling discovery that a square plate as well as a solid cube can be transversed by a continuously moving point in a finite time interval, although they are prototypes of geometric objects that no one calls curves.

Then there was the attempt to restrict the term 'curve' to *arcs*. An arc is the path of a point continuously moving during a finite time interval without assuming any position more than once. In other words, the sets obtainable from a straight line segment or an interval of real numbers by continuous one-to-one transformations.[9] Like a segment, each arc has two end points. In square-filling or cube-filling motions such as Peano's, some positions are traversed more than once: this can be proved to be inevitable. Thus squares and cubes are not arcs. But neither are circles and lemniscates, which everyone calls curves. While of great importance in itself, the concept of arc, therefore, is much too narrow for a definition of curve.

The realm of objects can be considerably widened if one unites several arcs. By an *ordinary curve* is meant a continuum that is the union of a finite number of arcs any two of which are either disjoint or intersect in a single point that is an endpoint of both arcs. Figure IV.3 shows a circle, a cross and a lemniscate, represented as such unions of 3, 4, and 6 arcs, respectively. But even this concept, which has important applications (e.g. to electrical networks), is too narrow to serve as a definition of the general idea of curves, since one can easily construct curves that cannot be decomposed into a finite number of arcs in any way.

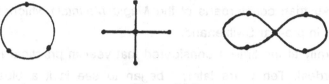

Figure IV.3

Cantor approached the problem from a different angle. Among the geometric objects in the plane, he defined curves as the continua that do not include any square domain, however small. This definition includes all plane objects called curves and excludes all not so designated. It thus is a perfect definition as far as it is applicable, that is, in the plane. But unfortunately it cannot be extended. In space, the surface of a sphere is a continuum that does not include any square domain, yet it is not called a curve.

These were the most important older attempts to define the concept of curve.

5.

The more I thought about these problems, the more they fascinated me and the more eager I became to complete their clarification. But after long studies in unheated libraries during the cold winter of 1921 and weakened by post-war malnutrition and overwork, I succumbed, in May 1921, to a serious respiratory infection, continuing to work feverishly in the literal as well as in the metaphorical sense. Too ill to attend the last meetings of the seminar I wrote Hahn a letter summarizing my results. But in the fall of 1921, my bout of the 'Viennese disease' became so serious that I felt it might be years before I could elaborate my ideas. After being bed-ridden for several months I had to retire for almost a year to a sanatorium in Styria — one of the two Austrian counterparts of the *Magic Mountain* which Thomas Mann set in pre-war Switzerland.

Naturally at the time, I considered that year in practical isolation as an ordeal. Ten years later, I began to see in it a blessing in disguise — not only because I emerged completely cured, but because my magic mountain had provided me with the quiet sur-rounding needed for contemplation, for reading great literature, for studying mathematics in some depth without oral help, and for developing my own ideas. These were rare opportunities for a young man even before the modern intensification of life. Looking back now, more than fifty years later, I think that even a conquerable disease was not too high a price. Unfortunately, in that pre-antibiotic era most of the victims lost the fight. At the time when I was recuperating on one of Austria's magic mountains, Franz Kafka was dying on the other one.

During the spring and summer of 1921, I elaborated the principle of my definition of curves — the consideration of the relation between a set and the frontiers of small neighborhoods of its points — in two ways.

First, it became clear that the basic idea lent itself also to a description of the *ramification* of a curve at its points. If the frontiers of some arbitrarily small neighborhoods of a point p contain at most 4 points of the curve C, then p can be said to be of a ramification of order at most 4. If the frontiers of all sufficiently small neighborhoods of p have at least 4 points in common with C, then p can be said to have a ramification order of at least 4. The cross point of a lem-niscate has an order at most 4 as well as at least 4, thus the *ramification order* 4. In the same way, one can of course define ramification order n for n = 1,2,3, . . . With the exception of the cross point, all points of the lemniscate and all points of a circle are of ramification order 2. The tripod (See Figure IV.4) contains one point of order 3, three points of order 1 (called *end points*), while all other points are of order 2. The distribution in a curve of the points of various ramification orders is one of the topics of what I began to call the *general theory of curves.*

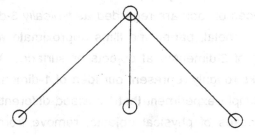

Figure IV.4

The second development started with the problem how to differentiate between the two principal types of continua in our space which are *not* curves: surfaces and solids. Traditionally this was achieved by stipulating that the points of a surface can be characterized by *pairs* of real numbers, while points in a solid require *triples* (just as in Descartes' analytic geometry points have two coordinates in the plane, and three in space). But while this description is adequate for most purposes of analysis it is quite unsatisfactory from a geometric point of view. For by the same token, each point of a curve should be characterized by a *single* number, as is a point of a straight line: yet this idea leads only to arcs or at best to ordinary curves. There is no way of describing each point on a general curve by a single number. Nor, for that matter, can one characterize a point on a general surface by two numbers.

I looked for clues in the physical world, even though nothing but well-defined geometric entities such as cubes, surfaces of spheres, and ellipses are unequivocally categorized as solids, surfaces, and curves. Material objects, for example an apple, its skin, and its stem, are so classified merely with a grain of allowance, for all material objects unless they are considered as dispersed sets of molecules are 3-dimensional. Yet only objects such as metal balls, wooden blocks and pieces of rock are regarded as *typically* 3-dimensional or as solids. Sheet metal, paper and films approximate what we mean when speaking of 2-dimensional objects or surfaces. Wire, threads, and chalk-marks roughly represent our idea of 1-dimensional objects or curves. A simple experiment that I devised differentiates between these three classes of physical objects: remove from an object a point and all points sufficiently near to it. The dimension of the object determines the *tool* required for the performance of the experiment. In the case of a wooden block, a *saw* is needed in order to cut the solid along surfaces. If the object is a cardboard surface, then a pair of *scissors* suffices, and the cuts are along curves. If the object is a wire curve, then all that is needed are *pliers* with which we have to pinch the curve in dispersed points. Finally, if the object is dispersed like a pile of sand, then *no tool at all* is required since nothing has to be dissected.

In the light of the definition of curves expounded above, the difference between surfaces and solids can be described as follows: Not being curves, a surface as well as a solid, S, contains a point, p, such that the frontiers of all sufficiently small neighborhoods of p have continua in common with S. It is the nature of these continua in the frontiers that accounts for the difference between the two types. If

S is a solid, then S penetrates the frontiers of all small neighborhoods of at least one point of S in entire *surfaces*. If S is a surface, then S cuts those frontiers along mere curves. A curve pierces those frontiers in dispersed points. Each point of a dispersed set is contained in arbitrarily small neighborhoods with whose frontiers the set has no point (or, in other words, the empty set) in common.

The preceding description descends from 3- to 2-dimensional sets, from 2- to 1-dimensional, from 1-dimensional to dispersed, from dispersed sets to the empty set. In order to continue the pattern it is convenient to call the dispersed sets 0-dimensional, and the empty set − 1 dimensional. Clearly, this is the end of the recursion.

On the other hand, in ascending from the empty set there is no need to halt at dimension 3 any more than one does in analytic geometry. There, one introduces the 1-dimensional Cartesian space (the line) as the set of all real numbers, the 2-dimensional space (the plane) as the set of all ordered pairs of numbers, the 3-dimensional space as the set of all ordered triples of numbers; and then one calls the set of all ordered quadruples of numbers the 4-dimensional Cartesian space, and so on. The physical space of our experience corresponds to a region in the 3-dimensional Cartesian space, as do mixtures of three substances and financial states resulting from three accounts. The space-time world of classical and relativistic mechanics, mixtures of four substances and financial states resulting from four accounts correspond to regions in the 4-dimensional Cartesian space; and Boltzmann's statistical mechanics made use of much higher dimensional spaces. For the past century, geometers have investigated objects in higher dimensional spaces for their own sake. There was no reason for me to halt at a particular dimension; and so I formulated the following general definition:

**A set S in a Cartesian space (of any dimension) is
n-dimensional if,**

**1) each point of S is contained in neighborhoods as
small as may be desired with whose frontiers S has
at most (n − 1)-dimensional intersections; and**

**2) for at least one point of S the frontier of each
sufficiently small neighborhood has at least a
(n − 1)-dimensional intersection,**

**[and, to start the process] the empty set [is
assigned] the dimension − 1.**

This definition, it will be noted, is free from any restriction to
continua or even to closed sets and it associates to each point set in
any Cartesian space a nonnegative integer — its dimension. It
represents, in other words, a classification of the immense variety of
all point sets. In particular, all geometric entities in the 3-dimensional
space, however complicated, ramified and convoluted, are divided
into five classes: the non-empty sets of dimensions 3, 2, 1, 0, and
the empty set of dimension − 1.

The enormous scope, indeed the all-comprehensiveness of the
dimension concept just quoted was at the root of some of the
epistemological interest of Schlick, Carnap, and Hahn, which was to
come up in discussions of the Circle.

Notes

[1]In order to avoid an unnecessary restriction of the studies to Euclidean space, Hahn
actually began by defining general metric spaces, which Maurice Fréchet had introduced in 1906. A metric space is a set of elements of unspecified nature, such that to
each pair of elements, a number is somehow associated, subject only to three
conditions. The elements of the set are referred to as *points*; the number associated

with the pair (p,q) is called the *distance* between the points p and q, and is denoted by d(p,q). But these terms are used only in order to evoke helpful mental pictures in the reader, while the mathematically essential part of the definition lies in the three conditions to which the distance is subjected:

1) d(p,q) is positive if p and q are different points, while d(p,p) = 0;

2) d(p,q) = d(q,p);

3) d(p,q) + d(q,r) is not less than d(p,r) for any three points, p, q, r.

Condition 3) generalizes the fact that in a Euclidean triangle the sum of two sides is not less than the third.

The Euclidean space is a metric space. So are the Euclidean plane and, as is easily verified, the taxicab plane mentioned in Chapter 3. The set of all real numbers is *not* a metric space, since if one associates to each pair (x,y) of numbers d(x,y) = y − x, conditions 1) and 2) are violated. For example, d(2,5) = −3, which is negative; and d(5,2) = 3, which is not equal to d(2,5). Nor is the set of all real numbers a metric space if one associates with each pair (x,y) the number d(y,x) = $(y - x)^2$, which satisfies Condition 1) and 2), but violates Condition 3); for example d(0,1/2) + d(1/2,1) = 1/4 + 1/4; and this is less than d(0,1), which is 1. But the set of all real numbers *is* a metric space if one associates to each pair (x,y) as distance the absolute value of y − x (that is, the positive square root of the number $(y - x)^2$), since this distance satisfies all three conditions.

Cantor's point set theory can be developed in general metric spaces exactly as in Euclidean space. A sphere of radius r about the center p consists of all points whose distance from q is strictly less than r. If the metric space is the Euclidean plane, then the sphere is the inside of a circle of radius r about q. In the taxicab plane, a sphere is the inside of a square circle.

[2]As can be easily shown, it is sufficient to demand that every sphere with the center p contain at least one point other than p.

[3]This stripping can be performed in thought only. When peeling an apple one cannot help but remove a solid rather than the mere surface. Scholastic philosophers speculated about the physical process of cutting a solid sphere into hemispheres. What happens to the points on the dividing circle? Are all of them incorporated into one of the hemispheres and if so, can one ascertain into which one? Or do some points go to one hemisphere, and others to the other? Or are some or all of the points

of the circle split? Reflection upon these problems suggests that matter is granular
rather than continuously distributed.

[4]Cantor had defined a continuum as a closed set (containing more than one point)
which, for any two of its points, p and q, and any positive number d, however small,
contains a finite chain of points beginning with p and ending with q in which the
distance between any two consecutive points is less than d. Jordan's concept can be
proved to be equivalent to Cantor's.

[5]Everyone using the term 'continuum' makes the following two assertions with equal
conviction: 1) The union of two disjoint segments is not a continuum. 2) Segments,
disks, and spheres are continua. But whereas 1) is an immediate consequence of
Jordan's definition of continua, 2) follows from that definition only by virtue of rather
profound theorems about real numbers.

[6]'Gordian' is the closest translation I can devise of Bieberbach's 'rattenköniggleich'
(rat-king-like) entanglement — a German metaphor that I have never understood.

[7]Even in that first conversation with Hahn I emphasized that in a curve C only the
frontiers of *some* small neighborhoods of each point p must be free of continua of C
(or, as I first suggested, have only a finite number of points in common with C), while
the frontiers of other small neighborhoods of p may well intersect C in continua. In the
following figure, the curve to be studied is the segment A.

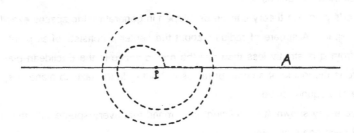

Figure IV.5

Two neighborhoods of the point p of A are suggested by their plane cross sections: a
sphere the dashed cross section of whose frontier intersects A in exactly two points;
and a snail-shaped neighborhood, the dashed cross section of whose frontier has a
single point and a segment in common with A. Clearly, p has arbitrarily small
neighborhoods of both kinds. If, by contrast, T is the part of the plane bounded by a

triangle, then the frontiers of *all* sufficiently small neighborhoods of a point of T do have continua in common with T. Hence T is not a curve.

There are curves of which even the frontiers of arbitrarily small *spheres* about a point intersect the curve in continua; for example, in a plane the union, B, of concentric circles of radii 1,1/2, 1/3,. . .,1/n. . . ., and a diameter of the first of these circles. Their common center, p, is a point contained in spheres of radii 1, 1/2. 1/3, . . ., 1/n, . . . , the frontier of each of which has a continuum (namely, a circle) in common with B. (One can easily extricate from B an arc with the same property.)

[8]Even the elements of *abstract* set theory became known to me (through Abraham Fraenkel's book on set theory) only after my conversation with Hahn.

[9]For many sets, as for example for segments and squares (but by no means all sets in Euclidean space), it can be shown that a continuous one-to-one transformation of S into a set S' also represents a continuous one-to-one transformation of S' into S. One-to-one transformations that are bicontinuous (that is, continuous in both directions) are called *topological*.

V. VIGNETTES OF THE MEMBERS OF THE CIRCLE IN 1927

1.

The meeting place of the Schlick Kreis was a rather dingy room on the ground floor of the building in the Boltzmanngasse that housed the mathematical and physical institutes of the university. The room was filled with rows of chairs and long tables, facing a blackboard. When we were not in session it was a reading room, occasionally used for lectures. Next door, there was a small library and a tiny office for Schlick and his assistant. Those who arrived first at the meeting of the Circle would shove some tables and chairs away from the blackboard, which most speakers used. In the space thus gained, they arranged chairs informally in a semicircle in front of the blackboard, leaving one long table for those who brought books along or wished to smoke or take notes.

People would stand in informal groups until Schlick clapped his hands. Then the conversations stopped, everyone took a seat, and Schlick, who usually sat at one end of the table near the blackboard, announced the topic of the paper or the report or the discussion of the evening.

The size of the Circle varied from 10 to 20 in the course of the years. During each academic year, the list of those attending remained on the whole the same, except for foreign guests.

The following vignettes are of the people who, to my recollection, attended the Circle when I joined in the fall of 1927.

2.

When I listened to some of Schlick's lectures as a student in 1923 and then took part in one of his seminars, he gave me the impression of being an extremely refined, somewhat introverted man. He was very sincere and unassuming, almost to the point of diffidence.

My admiration for his sincerity deepened as I came to know him more closely in later years. Empty phrases from his lips or the slightest trace of pompousness were unthinkable.

What caused the appearance of extreme modesty in his dealings with students, however, was Schlick's sometimes exaggerated politeness. While always prepared to correct his own views and to learn, he actually was perfectly self-assured. In fact, he was not free of sarcasm. If, in a discussion he heard someone utter an opinion that he did not deem to be quite up to his standards, for example, if the speaker said something in favor of parapsychological studies, he would almost imperceptibly shake his head, a more perceptible mocking smile would appear on his lips, and he would try to exchange glances with those who agreed with him, or who he thought were neutral. Persons so castigated sometimes ignored the punishment, but more often they would shorten their speech.

Schlick respected every person's political and religious convictions. When after Hitler's coming to power in Germany the Catholic philosopher Dietrich von Hildebrand followed a call of the Christian-

Social minister of education to the University of Vienna, and felt quite isolated in the faculty meetings of the then ultra-nationalistic School of Philosophy, it was Schlick who accompanied him and introduced him to less militant colleagues. But Schlick was strongly opposed to mixing intellectual studies with political ideals.

When I attended Schlick's lectures, I had already read his monograph 'Space and Time'[1] on relativity theory and parts of his main work, *Allgemeine Erkenntnistheorie* (General Theory of Knowledge), published in 1918.[2] As a lecturer Schlick was clear; but he always spoke in a low voice and usually in a monotone. While his speech was definitely not of the Austrian variety, neither was it characteristic of someone coming from Northern Germany. I had never inquired where he originally came. So once when I heard him at a party saying something about having been born in Berlin I asked with surprise "In Berlin?" "Yes," he said and half jokingly, half seriously added, "Sad, but true."

The first time I heard Schlick lecture I had some slight difficulty in understanding him because of his habit of breaking up some words into syllables — strangely, syllables in the sense of the English language, where one divides rid-er, rather than in the German sense of *Rei-ter*. But within an hour, I got completely used to this peculiarity of his.

In 1923/24 I took part in a seminar that Schlick conducted. Herbert Feigl and Friedrich Waismannn also attended. This was my first occasion to admire Schlick's talent for conducting stimulating discussions with calmness and unshakeable equanimity. Among other ideas, he brought up the elusive concept of the simplicity of a theory, a notion in which I was particularly interested because of an assertion repeatedly stressed by Henri Poincaré. While claiming that the physical world could be described by any consistent geometry, non-Euclidean as well as Euclidean, this great philosopher-mathematician felt that the Euclidean geometry would always be distinguished by the fact that it is *simpler than the other geometries* — *logically* simpler, Poincaré emphasized, in the same way as a linear polynomial is simpler than a quadratic polynomial.[3]

For Schlick's seminar, I stated some criteria for calling one theory simpler than others without, however, formulating a general definition of simplicity. Of course, Schlick had not expected me to give such a definition nor, for that matter, could he do it himself.

If, after close acquaintance with Schlick, a trace of the initial doubt about his self-assuredness remained, it was prompted by his inclination to idolize some people, if only figures of the first order. He studied physics under Max Planck and then, in the second decade of this century, upon becoming profoundly interested in the theory of relativity, venerated Einstein. There followed a period of deepest admiration for David Hilbert, during which Schlick himself, in his theory of knowledge, emphasized implicit definitions of concepts, i.e. the use of undefined terms satisfying postulates, which alone determine the meaning of those basic terms.[4] Then Schlick became fascinated with Russell. This was the state of things in 1927 when Schlick came under the ever increasing influence of Wittgenstein.

3.

Hans Hahn, in many ways the exact opposite of Schlick, was a strong, extroverted, highly articulate person who always spoke with a loud voice. In World War I he served in the Austrian army and was wounded on the Italian front. He was then appointed professor at the University of Bonn. But at the end of the war, he distributed pacifist leaflets — an act which, as his friend Schumpeter told me, made his position untenable in German universities. Hahn was a convinced socialist. In the faculty of the School of Philosophy at the University of Vienna, he always articulated his unpopular leftist convictions freely and forcefully. He protected the weak wherever he could.[5] In most other matters, however, he had a penchant for compromise — sometimes too great a penchant for my taste.

Scientifically, Hahn was a many-sided mathematician some of whose achievements have not been given due credit. For example, the normed vector spaces that mathematicians all over the world call Banach spaces were discovered independently and simultaneously by the Polish mathematician Stefan Banach and the Austrian Hahn and ought to be called Banach-Hahn spaces. Actually, the American Norbert Wiener also formulated the concept at the same time and independently of the two Europeans; but he never drew any inferences from his definition whereas Hahn, in the same way as Banach, used it as the starting point of a systematic theory. Besides, he is known for his work in the calculus of variations, on non-Archimedean systems, set-theoretical topology, real functions, and Fourier integrals.

Hahn had an extensive mathematical knowledge. Once at a meeting of German-speaking mathematicians, a group of the most outstanding of those attending in Hahn's age bracket tested their mathematical knowlege by asking each other questions, and Hahn was the winner of the tournament. He was a teacher of supreme clarity. In his lectures he proceeded by imperceptible steps, and at the end of each hour left his audience amazed at the mass of material he had covered. His first course as lecturer (*Privatdozent*) at the University of Vienna in 1907 dealt with the calculus of variations. It was attended, among others, by Erwin Schrödinger who, then close to his doctorate, kept detailed notes of Hahn's course.

The stimulating nature of Hahn's lectures and the clarity of his penetrating expositions is proved by the origin of my own interest in mathematics, as I have related in the preceding chapter.

A minor but appreciable part of Hahn's free time was devoted to parapsychological studies. He gave a few interesting public lectures on the subject, in which two points impressed me. One was taken from the famous French physiologist Charles Richet, who suggested that one should imagine a world in which all men, with the exception of a few scattered individuals, had completely lost the sense of smell. Walking with one of those few between two high stone walls, others might hear him say, "There are roses behind this wall." And to their amazement they would verify the statement. Opening some drawer he might say, "There is lavender in this drawer" and if none should be found he would insist, "Then there *was* lavender in this drawer." And sure enough, it would be established that two years earlier there was indeed some lavender in that drawer. The second point goes back to Hahn himself. Some mediums claim that great thinkers and poets speak through them while they are in a trance, but actually utter only lines that are far below the level of those writers. The usual interpretation is that the uneducated and fraudulent mediums simply say what they think those thinkers would say. Hahn, however, pointed out that many of those mediumistic revelations are indeed *so* trivial and incoherent that even a medium with little education would not consider them as utterances of those writers — in fact, they are definitely below the medium's *own* level. To Hahn, this fact indicated that such chatter was not a product of the conscious minds of the mediums, but was generated subconsciously. Its very triviality, combined with the tormented stammering in which the babble is frequently uttered, suggested to Hahn that in many cases one is dealing with a genuine phenomenon *of some kind*.

4.

Olga Hahn-Neurath, Hans Hahn's blind sister and Otto Neurath's wife, had that type of poise and calmness that one only finds in some of the blind. She attended the meetings of the Circle as well as her brother's seminars regularly, always smoking a big cigar. In her youth she had written three papers (one, jointly with her later husband) on the algebra of classes in Schröder's presentation,[6] which Clarence I. Lewis in his *Survey of Symbolic Logic* ranks "among the most important contributions to symbolic logic."[7]

5.

Otto Neurath was a man of immense energy and curiosity and was very fast in grasping new ideas. However, he looked at everything — ideas as well as facts — through an often distorting lens of socialist philosophy and with an eye to the possible effects of the ideas and facts on a socialization of society. I have never seen a scholar as consistently obsessed with an idea and an ideal as Neurath.

In conversations and discussions he was knowledgeable, lively, witty, and entertaining. He knew stories appropriate to almost any situation, usually eliciting a *se non è vero, è ben trovato* from his listeners. One that somehow illustrated the spirit of the Vienna Circle has stuck to my memory. He told it in a discussion of determinism when someone brought up the fable of Buridan's ass. Placed exactly midway between two exactly similar bundles of hay, the beast starved to death because it could not decide which bundle to turn to. "Listen to a more scientific version," Neurath said, and he told us of a starfish placed on the raised center of a tray, each arm in one of five symmetrically arranged cups underneath. Four cups were filled with water, one with nutritious solution; and soon the star fish slipped down into the right cup. When three adjacent cups were filled with a solution it took the animal much longer to decide and to slip slowly into the middle cup of the three. Finally all five cups were filled with the same solution; and the fish in the center starved and dried out.

In his youth, Neurath had written a few papers on Boolean algebra. In one of them, which he showed me but which I cannot locate any more, [8] he criticized the asymmetry of symbols such as p·q for symmetric operations or relations, since in our linear way of writing they give certain preference to p. He proposed to replace them by symbols such as

$$p \atop \cdot \atop q$$

In 1915, Neurath got intensely interested in optics through Mach's book on this subject and studied a classification of systems of hypotheses. In 1919 he served in the short-lived communist government in Munich, and was imprisoned for a short time when the regime broke down. He then fled to Vienna.

Later, Neurath wrote papers about the economic theory of value and at the height of the controversy about Oswald Spengler's *The Decline of the West*, he published a booklet 'Anti-Spengler.'[9]

Neurath was strongly opposed to any parapsychological investigations and considered all mediums frauds. He was particularly opposed to the involvement of Hahn (his brother-in-law), whom he did not regard as qualified to unmask the swindlers. "Who looks into these matters?" he once said to me answering his own question (of course in sociological categories), "Uncritical, run-down aristocrats and a few supercritical intellectuals such as Hahn. Studies of this

kind further the belief in supernatural forces and serve only reaction-
ary groups. And what results have ever come from this sort of
thing?" he added, repeating, knowingly or unknowing an argument of
Mach's. "Listen to the gibberish the mediums pass off as the words
of Goethe." I mentioned Hahn's reasoning that it is just the *complete*
nonsense produced by mediums that seems to indicate a genuine-
ness of some kind; but Neurath did not listen.

In the late 1920's, Neurath became greatly interested in contemporary graphic art; and in this field our interests met. He wanted to
utilize that kind of art for what he called *Bildstatistik* (Pictorial Statistics). In pictorial comparisons, say, of two nations, one often represents them by two men; and if one nation has twice the population of
the other, then it is represented by a man twice the size of the other.
Neurath naturally asked "But what is size? Is it height? area? volume?" The only unequivocal representation of two nations with populations of, say 80 million and 60 million is by two groups of 8 men
and 6 men (or, say, 4 men and 3 men or 16 and 12). This idea was
not really original. Such representations had been given before
Neurath. But he developed the method systematically and with great
taste. He invited excellent graphical artists, among them the German
Gerd Arntz and the Dutchman Peter Alma, — the Belgian Frans
Masereel he considered too romantic — and commissioned them to
design the units; e. g. the identical men representing the nations; the
ships standing for navies; the pictures representing units of produce,
industrial output. And he promoted this *Bildstatistik* with his full
enthusiasm and energy first in Vienna, later in Amsterdam, and finally
in London, where he founded the Isotype Institute.

Along with the great aesthetical enjoyment that I derived from the
work of the graphical artists of the 1920's, I also had the feeling that
their work was the artistic counterpart of the tendency to abstraction

in mathematics just as impressionism and the philosophy of Mach have been called artistic-scientific counterparts.

Abstraction began to pervade mathematics in the first decade of this century. Ernst Steinitz began to develop *algebra* in abstract fields rather than in the concrete fields of numbers that were a bequest of the 19th century. Fréchet generalized the idea of *space* by selecting only a few absolutely essential properties of the Euclidean, non-Euclidean, and some other concrete spaces and by defining an (abstract) metric space as anything possessing just those properties. While Fréchet and others, had used these abstract spaces in analysis (as I also did in the theories of curves and dimension), I found in the late 1920's that they can also serve as the basis of a *geometry* in the more traditional sense; and I developed what I called a *metric geometry*, including theories of betweenness, of geodesic line and curvature of curves and of surfaces in these abstract spaces. Analogously, the graphic artists of the 1920's selected only a few absolutely essential features of types of objects and people and represented, as it were, the Platonic ideas of those objects by these abstractions.

6.

Rudolf Carnap had visited Vienna in the spring of 1925 before I left the city for Amsterdam. At that time, he and I had a long talk about his work and about the theories of dimension and of curves which I was then developing and in which he was greatly interested since he had just finished a booklet about space.[10]

Carnap, especially in his younger years, had a face that almost seemed to exude sincerity and honesty. His manners were sedate, his motions slow. In whatever he undertook, he was thorough and systematic, sometimes to the point of pedantry. He was one of the most industrious scholars that I have ever met. He worked and learned constantly. When anything came up in conversation that was new to him or that he wanted to follow up, he would produce a little notebook and jot down a few words. As for parapsychological studies, he strongly favored the idea that they be carried on in a truly scientific spirit but was not inclined to participate in this work. He was not always fast in grasping ideas but he concentrated completely on whatever he was doing, saying, or hearing. For many years he refrained from driving a car, because as his wife told me, he could do only one thing at a time, and he knew that a driver must divide his attention.

Carnap, who, as far as I know, was a strict teetotaler, lived very modestly. All he wanted was the opportunity to do his research under reasonably comfortable conditions. Beyond that he had no material ambitions. He had socialist leanings but I had the impression that he never got around to thinking seriously about the economic problems connected with various forms of social organization. He was, however, a truly liberal and completely tolerant man, and stood up for the underprivileged and the persecuted whenever an occasion arose.

7.

Victor Kraft attended the meetings of the Circle regularly and often took notes though he rarely spoke in the discussions. He had started around 1911 as a philosophical realist in the sense of assuming the reality of an external world as a hypothesis for the explanation of the world of perceptions. He later became interested in Mach. In 1925 he published a long paper in the *Proceedings of the Vienna Academy of Science* (v. 203)[11] in which he to some extent anticipated

Popper's deductivism. In an abstract of this paper, Kraft said that "transmission of validity (*Geltungsvermittlung*) can only come about through rigorous deductive conclusions to which all methods, including induction, must be reduced. Induction, therefore, cannot truly prove general relations from specific facts. It can do so only on the basis of general assumptions, or as hypothetical generalizations that must continually be verified. Hence the clearest, most perfect construction of knowledge lies in a deductive axiomatic system, in which the hypotheses are explicitly formulated as initial assumptions while the development is purely deductive. Even the validity of such a system is only hypothetical since there are no immediately certain, self-evident propositions (at least about reality) which could serve as a starting point. Applied to the world of perceptions, the validity of such a theory consists in the agreement between the deduced and the observed facts." Kraft later applied these ideas to the *Geisteswissenschaften* (humanities), denying them *Verstehen* or intuition as an admissible special method.

Like Schlick, Feigl, and myself, Kraft by no means shared all the political ideas and ideals of Neurath, although the latter sometimes presented to the public, perhaps unintentionally, the idea of a politically homogeneous Circle. Schlick especially seemed to resent this since in Vienna, the circle was named after him, the *Schlick-Kreis*.

After World War II, Kraft wrote a book on the Vienna Circle.[12]

8.

Friedrich Waismann studied mathematics and philosophy at the University of Vienna in the early 1920's and took part in Schlick's seminars. I remember that in the seminar that I attended, he gave a good talk in which he sarcastically criticized a paper on geometry by

Oskar Becker, a student of Husserl's. He had always been a very clear expositor. Unfortunately Waismann dragged out his studies at the university. He seemed to be afraid of examinations which, being an unusually knowledgeable mathematician, he had not the slightest reason to fear. I believe that Schlick wanted to make him his official assistant; but a doctor's degree was a prerequisite for any academic position, however modest. So Schlick, Hahn, Carnap, and I repeatedly spoke to Waismann, encouraging him to get the formalities out of the way. But it became more and more apparent that Waismann had developed some block which made our attempts futile. This was the state of affairs in 1927 when he was already a very valuable member of the Schlick-Kreis, and just at that point he came under the spell of Wittgenstein.

9.

Herbert Feigl was another student of Schlick's. As I found out in many conversations with him during the winter of 1930 which we both spent at Harvard, Feigl had a limitless admiration for Carnap. This is the only explanation that I have for the fact, mentioned in Carnap's scientific autobiography, that Wittgenstein discontinued his relations with Feigl when he stopped seeing Carnap; for Feigl has always had an unusual ability to get along with everyone.[13] I have often felt it was a great loss to the Circle that in 1931 Feigl stayed on in the United States; and while he did probably more than anyone else to make some of the Viennese ideas known in America, in the subsequent discussions in the Circle, I badly missed him and the soberness that characterized his judgments.

10.

Theodor Radakovic had been an able student of Hahn's before the latter returned to Vienna. He was assistant professor of mathematics at the Polytechnical Institute in Vienna. I later induced him to collaborate with students of mine in developing lattice theory. He was too shy to take part in the discussions of the Circle although he attended the meetings regularly.

11.

Edgar Zilsel attended only a few of the meetings and, for reasons unknown to me, wanted to be considered only as close to, and not as a member of, the Circle. He was a very good lecturer and when he attended a meeting he usually presented an interesting paper. His main work was on the problem of application. However, I never read his book[14] since Hahn mentioned to me that at a crucial point a *non sequitur* invalidated the principal result of that otherwise good work.

Zilsel was a militant leftist. I once heard him give a brilliant talk at a meeting in Warsaw. I have forgotten the details but remember that he spoke about some philosophical views that could not, in his opinion, be justified on rational grounds. He ended, almost out-Neurathing Neurath, by blasting the socio-economic systems which he claimed, motivated those views and found support in them. I happened to be sitting next to the eminent logician Jan Lukasiewcz, who seemed to be simply overwhelmed by the talk and said repeatedly, "What an intellect!"

12.

Felix Kaufmann was a philosopher of law, a student of Kelsen's, and a member of the Kelsen, as well as of the Mises Circle. For a living, he worked as the Austrian representative of the oil company then called Anglo-Persian.

Being an ardent phenomenologist who cultivated ties with Husserl, Kaufmann did not wish to be counted as a member of the Circle; but he attended its meetings regularly. He was the only participant with a true sense of humor, Neurath's wit being always caustic. He wrote amusing satirical ditties in the Austrian-Bavarian dialect (called *Gstanzln*). One of his best *Gstanzln* had as background Spann's holist fight against traditional economics. It paraphrased a well known children's song by the Prussian Hoffmann von Fallersleben (who also wrote the words of 'Deutschland, Deutschland über Alles').

> The donkey and the cuckoo
> they quarrelled all the day
> who was the finest singer
> in the merry month of May.
> The donkey boldly shouted
> "It's me," with a loud bray.
> "But I can sing still better,"
> the cuckoo bird did say.
> The nightingale was questioned
> who should the winner be;
> and with a trill she answered,
> "It's all the same to me."

Kaufmann's *Gstanzl* I have ventured to translate as follows:

The whole with all its parts clash'd.
They quarrelled bitterly
who did in basic logic
deserve priority.
"It's we," a chorus shouted,
"The parts. There is no doubt;
and nowhere can you ever
find wholes that are without."
Then pompously the whole said
"It seems you all have missed
that only through my wholeness
do each of you exist."
A listening logician
said "This dispute's naive,
since neither one is prior:
you are correlative."

13.

All members of the Circle had a background of scientific research activity. Each of them took the scientific method seriously and in fact expected to obtain a consistent *Weltbild* (picture of the world) through what in the Circle was called *Die wissenschaftliche Weltauffassung* (the scientific apprehension of the world). Outside of the Circle, this attitude was highly praised by some, and severely condemned by others.

All this is by no means to say that science, even in the widest sense of the German word *Wissenschaft*, completely filled the lives of these philosophers. There was not a single aspect of the many-sided Viennese culture of that period that was not enjoyed by some members of the Circle. In particular, there was great interest in music, in dramatic art, in literature including lyric poetry, and even in some works of mystics.

But where cognitive content was concerned, they all fervently believed in clarity. Nothing was more odious to them than a hazy expression of presumed truths. Some expressions, as far as one could believe in the sincerity of the proponent, indicated to them at best his laziness or incompetence as expositor, at worst the imperfection of his thoughts. Metaphysical depth raised in them a strong instinctive distrust even before logical analysis revealed that it was not depth that characterized such speculations, but emptiness of cognitive content.

The members of the Circle considered the clarification of the basic propositions of their respective fields as the aim of their philosophical activities. There was little they could hope to learn for this purpose from the systems of traditional philosophy. But contrary to what has been said about them, they were not altogether ignorant of the history of philosophy.

Schlick was widely read in the philosophy of the ancient Greeks, of the Renaissance, and of the modern age. In particular, he had carefully studied Kant, the neo-Kantians, and Husserl. Kraft had more than the customary knowledge of a professor of philosophy. Carnap knew Kant, the neo-Kantians, and the Brentano school well. Hahn, who had edited Bolzano's *Paradoxes of the Infinite*, was a great admirer of Hume and of Leibniz, but had a strong dislike for Kant because of the frequent changes of meanings that the terms in the *Critique of Pure Reason* undergo. Neurath and Feigl, without being particularly interested in the history of philosophy, knew a good deal about it, especially the latter. Kaufmann, widely read, often looked at the topics under discussion in the Circle from a historical point of view and quoted the English empiricists, and — too often I thought — Kant. (Outside of the Circle it was customary in Vienna at that time to attribute to Kant all kinds of modern views, and to read them into the *Critique of Pure Reason*, while — likewise outside of the Circle! — I also heard a joke referring to what repelled Hahn in the *Critique*; that first of all, someone should translate Kant into German.)

It may not be inappropriate to mention in this connection, two mathematicians who attended the meetings of the Circle, Kurt Gödel (not listed among the members in 1927 because he joined the Circle only a little later), and myself.

My father owned a famous economics library with a comparatively small philosophical annex of about 1,500 volumes, which, however, included the collected works of practically all great philosophers from Bacon and Descartes to about 1900. In my last two pre-university years I made extensive use of this philosophical library: in fact, I wrote abstracts of the main works from Bacon to Fichte. When I started a serious study of physics and mathematics, however, I lost my interest in the traditional philosophical systems and, for a while, even looked on them with some aversion. Yet in 1922, a year after my father's death, when selling the economics library to the Hitsotsubashi University (then in Tokyo), I retained the philosophical annex, completed it during the next 10 years by acquiring the complete works of the few classical authors that had been missing (such as Pierre Gassendi and Pierre Bayle), and brought it up to date especially along the lines of logic and philosophy of science, which interested me more and more.

Gödel studied a great deal of philosophy. In particular, he had always been most intensely interested in Leibniz. Like myself, he also read Kant carefully; but whereas my own reading of Kant's

German successors had never gone beyond Fichte, Gödel read more. One day he showed me from one of Hegel's works a perfectly amazing passage that might have been a quotation from Einstein. Although Hegel could not possibly have written it in the spirit in which a scientist reads it today, Hegel seemed in that passage to anticipate the general theory of relativity.

All in all, it seems fair to say that the knowledge of philosophy in the Vienna Circle was not confined to the works of Mach and Russell, as some opponents have intimated.

Notes

[1]'Raum und Zeit in der gegenwärtigen Physik. Zur Einführung in das Verständnis der allgemeinen Relativitätstheorie', *Die Naturwissenschaften*, 5, 161-167, 177-186.

[2]Mortiz Schlick, *Allgemeine Erkenntnislehre*, Berlin: Springer, 1925, and *General Theory of Knowledge*, trans. by Albert E. Blumberg, Vienna-New York, Springer, 1974.

[3]Little did I then suspect that 15 years later the concept of simplicity would be as elusive as ever but that I myself would question the thesis of Poincaré concerning the simplicity of Euclidean geometry. I would challenge it by developing Janos Bolyai's and Nicolai Lobachevsky's non-Euclidean geometry from fewer basic concepts than are demonstrably indispenable for the development of Euclidean geometry.

[4]Actually, however, while Hilbert contributed immensely to this idea and vigorously propagated it, its originator, who was also the founder of the axiomatic method, the geometer Moritz Pasch clearly developed it fifteen years before Hilbert but never received proper credit for this truly fundamental achievement. He remained at the relatively minor University of Giessen for the rest of his life.

[5]Once he saw a man torment a horse and when his pleas were ignored he dragged the ruffian to the police.

[6]Olga Hahn and Otto Neurath, 'Zum Dualismus in der Logik', *Archiv für systematische Philosophie* 15, (1909) 149-76.

[7]C. I. Lewis *Survey of Symbolic Logic*, reprint (1960) p. 211

[8][The editors have not been able to find a published paper to this effect.]

[9]Otto Neurath, 'Anti-Spengler,' Munich, Georg D.W. Callwey, 1921, 96p. Republished

in *Empiricism and Sociology*, Dordrecht, D. Reidel, pp 158-213.

[10]*Der Raum. Ein Beitrag zur Wissenschaftslehre.* 'Kant-Studien' Ergänzungshefte, Nr. 56. Berlin: Verlag von Reichard, 1922, 87 pp.

[11]Victor Kraft, 'Die Grundformen der wissenschaftlichen Methoden', *Sitzungsberichte der österreichischen Akademie der Wissenschaften, philos.-histor. Klasse* 203, 1-104 (second edition 1973).

[12]*Der Wiener Kreis. Der Ursprung des Neopositivismus. Ein Kapitel der jüngsten Philosophiegeschichte*, Vienna: Springer, 1950, and in English trans, Arthur Pap, Vienna, New York, Springer, 1968.

[13]Rudolf Carnap 'Intellectual Autobiography', *The Philosophy of Rudolf Carnap*, Paul Arthur Schilpp, ed., LaSalle, Open Court, 1963, p. 27, "From the beginning of 1929 on, Wittgenstein wished to meet only with Schlick and Waismann, no longer with me or Feigl, who had also become acquainted with him in the meantime, let alone with the Circle."[In fact, perhaps for a variety of reasons, throughout 1928 Wittgenstein was seeing Feigl but not Carnap. B.McG.]

[14]Edgar Zilsel, *Das Anwendungsproblem; ein philosophischer Versuch über das Gesetz der grossen Zahlen und die Induktion*, Leipzig, J. A. Barth, 1916.

VI. REMINISCENCES OF THE WITTGENSTEIN FAMILY

1.

I record here what little I know about the remarkable, widely ramified Wittgenstein family because it may give a more intimate understanding of the milieu in which the philosopher grew up than do those of his biographies that have come to my attention. These biographies stress that Ludwig Wittgenstein came from a rich Jewish family in Vienna. This emphasis is liable to portray an incorrect picture. I knew quite a number of wealthy people belonging to the cutural élite of Vienna; and in so far as there were differences between Jewish and non-Jewish families I must say that the part of the Wittgenstein family that I knew resembled the latter.

2.

The Wittgensteins belonged to the wealthiest group in Austria. Before World War I, their combined fortune had been estimated at 200 million crowns — the equivalent of at least that many dollars after World War II. Their wealth originated in the steel industry of the Austrian empire, largely developed by Ludwig's father and grand-father, who were the Carnegies of Central Europe.[1]

This grandfather, Hermann Wittgenstein, who died in 1876, was a man of the highest cultural standards. He arranged musical evenings for his family and the circle of his friends, which included lumina of the famous medical school. By marriage, the Wittgensteins were related to the family of the great physiologist Ernst von Brücke and to the Kupelwiesers, who before World War I developed the beautiful Adriatic Island of Brioni. Brahms was a frequent guest and played the piano.

Fifty years later, Hermann's daughter Clara told me that when she was a young girl her father took her and one of her sisters (later married to government councilor and professor Oser) to Holland. There, in order to acquaint them with the Dutch language and culture, he engaged as a tutor a promising young writer who proposed to read with them the manuscript of a novel he was about to publish. The young man, Frederik van Eeden, became one of Holland's renowned authors; and that novel, *De kleine Johannes*, made him famous. ("The instruction was only moderately successful, though," the old lady added smiling, "Since neither of us ever succeeded in pronouncing those three Dutch words to van Eeden's satisfaction, we never went beyond the book's title.")

In the first decade of this century, the family was known in Vienna for their keen awareness of the social obligations of wealth. In particular, two of Hermann Wittgenstein's children, Clara and Karl (Ludwig's father) donated millions to charity and generously subsidized, among others, a Viennese organization whose aim it was to aid people in financial distress. An aunt of mine (a sister of my mother)[2] had been employed by this organization as a social worker since the first decade of this century. Her work impressed Clara Wittgenstein, in whose house she met Lydia Oser, an unmarried niece of Clara, who had difficulties in walking because of a crippling

childhood disease. A close friendship developed between my aunt
and Miss Oser, who despite her handicap became active as a social
worker.

3.

Miss Oser was the first member of the Wittgenstein family that I met.
Even then, as a young boy, I was deeply impressed by her totally
unassuming, almost self-effacing personality. She was kindness and
thoughtfulness personified.

In the summer of 1923, while roving through the Alps, I came to
Thumersbach on the beautiful Zeller See in the province of Salzburg,
where Miss Oser was spending a vacation on an estate of her aunt
Clara. I dropped in to see her, was cordially received, and met Miss
Clara Wittgenstein, then almost 70. She was a frail looking person of
the utmost simplicity; but a conversation with her soon revealed not
only a superior mind, extensive knowledge, and wide interests but
also a very strong will. There was in her speech that trace of
Viennese dialect that was characteristic of the German then spoken
in the circles of the highest bureaucracy and by the Austrian aristocrats. Clara's numerous nieces and nephews — especially, as I
was told, Ludwig Wittgenstein — revered her deeply.

After that visit to Thumersbach, Miss Wittgenstein, invited me
from time to time to her home in Vienna and to her estate in
Laxenburg, a baroque palace built in that suburb by the empress
Maria Theresa for one of her generals or ministers. The rooms with
their enormous ceramic stoves might have served as models for the
scenery of Der Rosenkavalier. When Clara's parents were still alive it
was there that Brahms played the piano. There, Clara also told me, a
letter was found in which Brahms called a close relative of hers (I

have since forgotten whom) an extraordinarily gifted musician and encouraged his choosing music as a career. By some inexplicable mistake, the unopened letter remained misplaced for many years and the Wittgenstein commended never learned what Brahms, whom he idolized, had written about him. "It might have changed the course of his life if he had," she added.

A strong musical tradition lived on in the family. Clara Wittgenstein had a string quartet under her patronage and arranged musical afternoons in her Viennese home. On two of these occasions I briefly met her nephew, Ludwig's brother, the pianist Paul Wittgenstein, who had lost his right arm during World War I, I believe in Russia. Despite this terrible handicap he decided to continue his career as a concert pianist. One of his main problems was the paucity of piano pieces for the left hand; and of course he did not want to repeat these same few pieces over and over again. So he commissioned concertos for piano left-hand from some of the greatest composers of the first quarter of the century including Ravel, Reger, and Richard Strauss. These pieces he then played in public concerts.[3]

At the dinners in her home, Clara spoke in the liveliest way of the latest events and books. The old lady was one of the first in Vienna to read Sinclair Lewis, whose *Arrowsmith* greatly fascinated her. Another time she discussed the early photos of David Octavius Hill, collected in a book by Heinrich Schwartz, commented on the expressiveness and beauty of British heads, and sent me a copy of the book. Only once in our many conversations over the years did I discern anything that seemed to me to suggest her age. When I gave her Frans Masereel's *Le Soleil*, a novel in woodcuts, which aroused the enthusiasm of young people and even the admiration of Thomas Mann, Hermann Hesse, and some men of their generation, she seemed to praise this ingenious work only out of politeness. But she took my part when I defended the artistic qualities of Eisenstein's *Potemkin* in a heated discussion with another (much older) guest. In a subsequent year, before my first trip to the United States, she drew to my attention, and sent me a copy of, the delightful booklet that Boltzmann wrote upon his return from America, *A Professor's Journey to Eldorado*.[4]

These are just a few samples from the wide range of her interests. Her mode of life and that of her relatives was, contrary to Thorstein Veblen's description of the ways of the rich, characterized by the unconspicuous consumption of goods of the highest cultural value.

In Clara's house, I met several relatives of hers, in particular, on a number of occasions her niece Mrs. Margaret Stonborough-Wittgenstein, one of the sisters of Ludwig and Paul, married to an American but living with him in Austria. When I first saw her entering through a door, she seemed to step out of the frame of the portrait of her that Klimt had painted twenty years earlier, when she was a young girl, but in which he prophetically showed her likeness as a mature woman.

While the other members of her family lived quiet, rather retired lives, the Stonboroughs were frequently seen at official functions. After World War I, they donated funds to the Viennese Academy of Science, enabling that institution to continue the publication of its proceedings. Mrs. Stonborough spoke with great respect, almost admiration, of the Viennese minister of finance, Breitner (see I.1), although, as I had been told, her lawyers had filed law suits against the city in taxation matters.

Mrs. Stonborough was a very intelligent person with various interests that included mathematics. In this field she had been tutored by a Professor Rothe of the Vienna Polytechnic, whom she often mentioned, always with respect and gratitude. Once, I remember, I took a slip of paper, jotted down the words

send

<u>more</u>

money

and told her of the problem, (then new in Vienna) of replacing letters by digits in such a way that a correct addition results. She looked at the paper for not more than two seconds and said, "Well, m = 1," which is of course the key to the solution.

In 1927 just after the house that her brother Ludwig had designed and built for Mrs. Stonborough was completed, she invited me there for dinner. The building stood on top of a little knoll in a garden that took up a small city block located in a not very attractive part of Vienna. The garden was surrounded by a high wall so that from the street one could hardly see the building. The house was uncompromisingly modern, without, however, resembling the work of Loos, whom Wittgenstein knew well. The beautiful interior offered some impressive vistas. There was a well lit hall from which one could look into several rooms. Its only decoration was a Greek statue of white marble in a niche.

At the dinner the conservation turned to Brecht-Weil's *Dreigroschenoper* (*Threepenny Opera*) which for some months had created a sensation in Vienna. I commented on the cleverness of some of Brecht's lines but received a disappointed and disapproving glance from the hostess. It turned out that she and some of her guests considered the work superficial since it did not treat social problems in the spirit of ethics. I believe that at that time Mrs.

Stonborough was a spiritual follower of Tolstoy, as were at various periods several members of her family including her brother Ludwig.

Present at the dinner party were, among others, one of the sons of the hostess and, if I remember correctly, two adopted sons from Germany. The young men did extensive social work for prisoners. But for all her social conscience, Mrs. Stonborough seemed to me to belong to that type of very rich Europeans who consider important positions even more than wealth as a birthright of their children. "They must become reformers of some kind," she told me one day. "That is the only career befitting our family." And another time she said, "I want to be remembered as the daughter of my father, the sister of my brothers, and the mother of my sons."

4.

A sister of Lydia was married to the noted German professor of education, Hermann Nohl. Carnap, one of the prominent members of the Vienna Circle, studied at the University of Jena before World War I and held Nohl in high esteem. Carnap's main teacher in Jena, however, was Frege, the father of the enormous development of logic after Boole (see VIII,4,5). Now, half a century after his death, Frege is revered by all logicians the world over. But around 1910 he was completely unknown, and Wittgenstein was one of his very first admirers.

In the mid 1920's I once met Nohl in Lydia's house and suggested that having spent several years in Jena he might have known Frege. "Of course," was the answer, "Who didn't?" "He certainly was one of the great logicians of all time," I remarked. "Actually most people laughed at Frege and considered him rather foolish," Nohl said. "But one day, a rumor circulated that, after all, some people abroad must be even more foolish since one man had just come to Jena all the way from England in order to visit Frege." (1908 was the year when Frege received the visit of Bertrand Russell, the real discover of Frege; and Ludwig Wittgenstein came in 1912.)[5]

5.

In 1930, I accepted an invitation for a year in the United States. After my departure I never saw Mrs. Stonborough again.[6] My relations with Clara Wittgenstein continued until her terminal illness in 1935. Lydia Oser-Wittgenstein, whom her aunt Clara had legally adopted in that year, spent the last years of her life on the estate in Thumersbach. Visits of her family, especially her cousin Paul, who played the piano for her, were among the highlights of her life, and that of my aunt, who was Lydia's guest from the beginning of the war to her death in the mid 1950's.

Soon after the war I asked my aunt to send me a little paperback edition of Peter Hebel's *Schatzkästlein* (*Little Treasure Chest*), a collection of old-fashioned stories some of which I had enjoyed as a boy and now wanted to read to my children. My letter was answered by Lydia, who said it might interest me that the *Schatzkästlein* was also a favorite book of Ludwig Wittgenstein's. He once discussed it with Clara and then gave her a copy of the book. Lydia had inherited this copy and was sending it to me as a souvenir.

In the early 1960's, when I finally overcame my aversion to seeing Austria again, I visited Lydia in Thumersbach on each of my trips. She died in the mid 1960's.

Notes

[1][Ludwig's father was indeed a steel magnate. The grandfather's lesser, though still

considerable, fortune was derived from agricultural, real estate, and other business interests. B. McG.]

[2][Emilie Glaser, A. S.]

[3][Paul Wittgenstein's life is described in E. Fred Flindell, "Paul Wittgenstein, patron and pianist,"*Musical Review* xxxii (1971) 107. B. McG.]

[4]Ludwig Boltzmann, 'Reise eines deutschen Professors ins Eldorado', *Populäre Schriften* 22 1905, pp 403-435.

[5][Russell's correspondence with Frege began in 1902. Apart from this remark of Menger's I can find no evidence of a visit in 1908, or indeed at any time. For Wittgenstein's visit, see my own *Wittgenstein a Life: Young Ludwig*, London & Berkeley CA, pp. 82-84 B. McG.]

[6][Yet Mrs. Stonborough remained in Vienna and Gmunden except for the war years, which she spent on New York. B. McG.]

VII. LUDWIG WITTGENSTEIN'S AUSTRIAN DICTIONARY

1.

Upon his return from World War I in which Wittgenstein had served in the Austrian army, he gave most of his property away — mainly, I have been told, to his sisters and his brother — and decided to become an elementary school teacher.[1] He taught for six years altogether in three communities in the province of Lower Austria: in the Alpine village of Trattenbach, about 60 miles south of Vienna; in the somewhat larger community of Puchberg at the foot of the Schneeberg, a 6000 ft. peak in the Eastern Alps; and in Ottertal. In letters to Bertrand Russell in the early 1920's, Wittgenstein bitterly complained about the 'wickedness' of the people in one of these places. During that period he published a short dictionary for the use of pupils in Alpine elementary schools. Except for a short paper, this dictionary was to be Wittgenstein's only publication during his life time apart from the *Tractatus Logico-Philosophicus*. I had wanted to see the booklet ever since I heard about it. But it has been out of print since the 1920's.

In 1961 when I visited Lydia Oser-Wittgenstein for the first time after World War II, she mentioned her cousin Ludwig, which she had never done in our previous conversations. "Is there really any merit," she asked, "in saying 'Whereof one cannot speak, thereof one must be silent'?" I said that in my opinion this last sentence of the *Tractatus* was of great importance for philosophy, in fact, that I regarded it as one of the most fundamental of all of Wittgenstein's sayings; and I tried to explain its significance.

I took the opportunity to ask Lydia where I might get hold of a copy of Wittgenstein's dictionary. She wrote to Mr. R. Koder in Vienna, who had been a colleague and friend of Ludwig's during his stay in Puchberg, and asked him to give me the desired information. Mr. Koder then kindly lent me the copy that he had used as a teacher in the 1920's.

2.

The dictionary[2] is a pamphlet of 42 pages. It contains approximately 5,700 entries including 110 Christian names and 36 geographical names. Fewer than 10% of the entries are followed by very short explanations many of which consist of a single word. The other 90% of the entries are simply listed.

In a letter, Mr. Koder explained, that before the booklet was published "there existed only an antiquated official dictionary, which did not include words in daily use such as 'to go' but did mention countless farfetched foreign words. This made the use of the book very difficult for the youngsters. That was why Wittgenstein published a dictionary that contained all German words necessary for pupils in elementary schools and only the most common foreign words. Moreover he wished to further the study of correct spelling and grammar by quoting examples of the Alpine vernacular and by the use of diverse typography. For example, he stressed (by using bold face type for the terminal letters) the differences between '*das*' and '*dass*',

and between the dative '*ihm*' and the accusative '*ihn*', (often confused in the dialect). When the booklet was being used it became
obvious that some important words were missing. Wittgenstein intended to insert those words in a second edition which, however, did
not materialize since the book did not achieve the acclaim that it
deserved and because Wittgenstein gave up teaching and moved to
Cambridge."

3.

A glance into the booklet convinced me that the dictionary certainly
fulfilled one of the aims mentioned by Mr. Koder. It is easily readable
and free of farfetched words. The few foreign words in it, e.g.
Bibliothek, *Coupé*, *Friseur* (for library, compartment, barber) and
Greco-Latin names of sciences, are used in the everyday language
of Austria.

The concessions to the Austrian vernacular are undoubtedly
useful in Alpine elementary schools. Besides the features mentioned
by Mr. Koder, one notices a few typically Austrian words, not understood in Germany, such as *Obers*, *Haferl* and *Spagat* (for cream,
little pot, and string), as well as characteristically Austrian abbreviation of some Christian names, among them *Nazi* for Ignatius (long
before the word acquired its current meaning as the abbreviation of
National Socialist).

4.

I did not detect any system in the selection of entries. Besides
Vienna, only two cities are included: New York and Prague. Of the
eight Austrian provinces four (among them Tyrol) are listed, the

others (among them Salzburg) are missing. Sweden is the only Scandinavian country included. Only *Amerika* appears in the list, but no reference to the United States. 'Russia' is missing altogether, so are some rather common Christian names, among them the author's very own, 'Ludwig.' Of the sciences, the dictionary includes astronomy, chemistry, and mechanics, but not physics. The missing words are probably among the gaps that were to be filled in a second edition. But in many cases it is hard to see how they could be overlooked initially.

The selection of the 10% of the words which are not only listed but explained appears at first glance to be altogether haphazard. The choice of many of these words, however, is well motivated, since the intention is evidently to teach the unlike spelling of words that sound alike or almost alike (such as the German words '*sinken*' and '*singen*', '*hohl*', and '*holen*'). Analogously, in an elementary English dictionary one might explain 'sink' by the definition 'drop slowly' and let 'sing' be followed by the words 'a song.' But there are also many cases that have nothing to do with phonetics; and in these, the selection of the words followed by explanations really is unsystematic and haphazard. For example, '*Resi*' (which is an abbreviation of 'Theresa') is followed by the explanation '*von Therese*' (from Theresa), whereas 'Mina' (which is an analogous abbreviation of 'Hermine') is merely listed and not explained. 'Mars' and 'Jupiter' are explained by the word 'planet', whereas 'Venus' and 'Saturn' are listed without an explanation; ('Mercury' is not listed at all). The pronounciation of '*Chauffeur*' is explained, that of '*Bureau*' is not.

As in the aforementioned English example of 'sink' and 'sing', Wittgenstein uses various forms of explanation: in some cases definitions; in others, association; in still others, subsumptions. This is reasonable. If one wants to make clear what English word is spelled 'sing' in contrast to 'sink', the remark 'a song' is perfectly conclusive and yet much shorter than any definition of 'sing' could possibly be, whereas 'drop slowly' is a fairly satisfactory and short definition of 'sink.'

The connection of entries with the explanations following them is established by four different methods: an equality sign; a semicolon; parentheses around the explanation; and the word 'von' (German for 'from') in cases of derivations. But the application of these methods is totally unsystematic. Nor are the connections correlated with the aforementioned different forms of explanations. For example, among the (typically Austrian) abbreviations of Christian names, one finds

Sepp = Joseph; Mitzi (Mary); Resi (from Theresa);

(Here and in what follows, the German entries in the book are translated into English.)

Other examples include:

fin = fin of a fish; bark : bark of a tree; archer (hunter); steamer = steamship; velvet : fabric; hatred (from hate); hollow : empty; hovel (hut).

An English analogue to the unlike treatment of homonyms in the booklet is

top = mountain top; top : toy.

In some cases, the use of the equality sign is at variance with all customary usage; e.g. in

adder = snake.

As every youngster in the Alps knows, there are several kinds of snakes besides adders.

5.

Criticism of this dictionary cannot help being pedantic; but for all their pettiness, the preceding remarks seem to make it hard to escape the inference that, while the booklet was of some limited practical usefulness in the region for which it was destined, it was not the work of a systematic mind.

Notes

[1][For background to this, see my *Young Ludwig* (already cited) pp. 277-82. B. McG.]

[2]Ludwig Wittgenstein, *Wörterbuch für Volksschulen*, Vienna, Hölder-Pichler-Tempsky 1926, [reprinted, though Menger seemed unaware of it, 1977, B. McG.]

VIII. WITTGENSTEIN'S TRACTATUS AND THE EARLY CIRCLE

1.

Ludwig Wittgenstein had enough first-rate ideas to influence a variety of thinkers; he expressed some ideas vaguely enough to keep hosts of interpreters busy; he changed them often enough to provide work for some score of biographers and historians; and he shrouded them (and himself), in enough mystery to originate a cult.

Each of these facets, which have made Wittgenstein the most widely discussed philosopher of this century, at least in the English-speaking world, played at one time or other, a role in the Vienna Circle. In this chapter, I shall try to reconstruct the influence of the *Tractatus Logico-Philosophicus* on the Circle in the years before I attended its meetings. It was in 1924, I believe, that the geometer Kurt Reidemeister, a member of the early Circle, studied the book at Schlick's and Hahn's request and presented an extensive report about it in a meeting.

2.

Three aspects of the *Tractatus* were perfectly consonant with the
spirit that prevailed among the members of the Circle even before
they had heard about the book: Wittgenstein's *antimetaphysical*
attitude; his *positivism*; and his view that philosophy is not a theory
yielding 'philosophical sentences' but rather *an activity aiming at the
logical clarification of thoughts*.

The third idea is formulated in Tr. 4.112.[1] Wittgenstein's
antimetaphysical attitude is apparent in the observation that "most
propositions and questions that have been written about philosophical matters are not false but senseless"(Tr. 4.003). His positivism is
manifest in his description of the right method in philosophy, which
begins with the stipulation (Tr. 6.53) that one should "say nothing
except what can be said, that is, propositions of natural science" and
in the sentence that "the totality of all true propositions is the total
natural science (or the totality of the natural sciences)"(Tr. 4.11; see
also III,7).

The clarifying rather than dogmatic role of philosophy is also
clearly described in Schlick's book *Allgemeine Erkenntnislehre*
(*General Theory of Knowledge*, 1918). Moreover, Schlick emphasized this role of philosophy in his lectures even before he had seen
the *Tractatus*. This concept was probably also dormant in the minds
of some of the other members of the Circle. It certainly was strengthened in all of them by their study of the *Tractatus*.

But Wittgenstein's antimetaphysical and positivistic attitude is by
no means unequivocal. This is manifested by some of the closing
statements in the *Tractatus* itself. Several members of the early
Circle tried to ignore those passages. But Neurath, the Argus of the
Circle, repeatedly quoted, "The sense of the world must lie outside

of the world"(Tr. 6.41). "Not *how* the world is, the mystical, but *that* it is"(Tr.6.44), "There is indeed the inexpressible. This *evinces* itself (dies *zeigt* sich); it is the mystical"(Tr.6.522). These and some other remarks created in him a deep distrust in Wittgenstein's philosophical standpoint.

Other aphorisms in the *Tractatus* supplied some members of the Circle with themes, often followed by variations, in their conversations, especially the following two: "In order to recognize the symbol in the sign we must consider the significant use"(Tr. 3.326) and, "To understand a proposition means to know what is the case if it is true"(Tr.4.024).

3.

Three other aspects of the *Tractatus* came as truly great revelations to the original members of the Circle, who had started from pure empiricism, but were dissatisfied with Mill's and Mach's views on logic and mathematics. These were *the stress on a critique of language; an attempt to clarify the role of logic*; and *the definition of, and emphasis on tautologies*.

Stress on language is evidenced, e.g. in the contention that "Most of the questions and propositions of the philosophers result from the fact that we do not understand the logic of language,"and "All philosophy is 'critique of language'."(Tr. 4.003 and Tr. 4.0031).

Wittgenstein's emphasis on the critique of language had to a large extent been anticipated by Fritz Mauthner at the turn of the century (see III,4). Wittgenstein gave him some backhanded credit by adding to the sentence that all philosophy is 'critique of language' the parenthetical remark: (*"allerdings nicht in Sinne von Mauthner"*). This, correctly translated (as it has been in the second translation of the *Tractatus* by D. F. Pears and B. F. McGuinness, 1961), means "though not in the sense of Mauthner." But for four decades most English-speaking readers probably relied on the inaccurate first translation "not at all in the sense of Mauthner,"which must have discouraged readers from studying Mauthner still more than Wittgenstein himself had done in the original German version.[2] True, Mauthner hardly knew mathematical logic and certainly did not clarify the role of logic and of tautologies, except for insisting (in anticipation of Wittgenstein) that they do not supply information about the world. On the other hand, Mauthner's critique of language could well apply to certain phases of Wittgenstein's own work (see VII,2 and IX,7).

As to the nature of logic, Hahn and probably most of the other members of the Circle originally agreed with Russell's view according to which a logical proposition is characterized by the fact that it is made up of connectives such as 'and', 'or', 'not' (called logical *constants*) and *variables* in the sense of 'symbols whose meaning is not determined' or which 'represent *any object for which the proposition is either true or false.*' A nonlogical proposition contains also names of individuals or references to definite objects in the world.

Wittgenstein's starts with propositions compounded out of others by logical connectives. While objects can be represented by signs, "the logical constants do not represent"(Tr. 4.0312). The truth (T) and falsity (F) of a compound depends exclusively upon the truth and falsity of its components. This is illustrated in the following 'truth tables' for compounds of two propositions (indicated by p and q) by the connectives 'and', 'or'[3] and 'implies', and for the 'compound' of one proposition with the particle 'not'.

p	q	p and q	p or q	p implies q
T	T	T	T	T
T	F	F	T	F
F	T	F	T	T
F	F	F	F	T

p	not p
T	F
F	T

The truth tables for the propositions 1) 'p and not p', 2) 'p or not p' can be obtained by use of the third and fourth lines of the table for 1) 'p and q', 2) 'p or q', respectively, with the substitution of not p for q.

p	not p	p and not p	p or not p
T	F	F	T
F	T	F	T

A compound proposition that is true in all cases (i.e. regardless of the truth or falsity of its components) is what Wittgenstein calls a *tautology*; a compound that is false in all cases he calls a *contradiction*. Examples are 'p or not p' and 'p and not p', respectively. But 'Socrates is mortal or Socrates is not mortal' and 'it is raining or it is not raining' are also tautologies, notwithstanding their references to Socrates or the weather. No tautology says anything about the world or in fact anything at all (Tr. 5.43). Wittgenstein's view of logic can be epitomized in the sentence: *The propositions of logic are the tautologies, and they say nothing.*

Truth tables were introduced by Charles Sanders Peirce in the early 1880's in some profound papers in the *American Journal of Mathematics* which, however, attracted little attention on either side of the Atlantic. Truth tables were rediscovered and tautologies discovered, simultaneously with and independently of Wittgenstein, by Emil L. Post, one of the greatest logicians between the world wars. In his doctoral thesis,[4] which was unknown to the early Circle, Post introduced tautologies and contradictions under the less suggestive names of *positive* and *negative functions*.

4.

Post's thesis contained, moreover, an important characterization of tautologies, which is independent of truth tables: *The tautologies are just those propositions that can be derived from Frege's axioms.* Since extensions of this theorem were to play an enormous role in Vienna, this section digresses into an explanation of Post's result. But nothing in the sequel depends upon the material in this section.

Frege was the first to axiomatize the calculus of propositions. He started out by admitting as axioms six propositions in terms of implication and negation, later reduced by Lukasiewicz[5] to the following three:

(1) p implies (q implies p);

(2) (not p implies not q) implies (q implies p);

(3) (r implies (p implies q)) implies ((r implies p) implies (r implies q)).

Frege furthermore admitted all propositions derived from the axioms or from their consequences by the following two procedures:

(I) *Modus Ponens*. If the propositions 'p' and 'p implies q' are admitted, then q is admitted.

(II) *Substitution*. If in an admitted compound a letter, say p, is replaced by another letter or by any compound — by the same one for all occurrences of p — then the resulting proposition is admitted.

For example, if in (1) p is replaced by r one obtains

r implies (q implies r);

replacing q by p yields

r implies (p implies r);

replacing here r by q yields

(1*) q implies (p implies q).

Being obtained from Postulate (1) by repeated application of Procedure (II) the formula (1*) is admitted. It could be obtained from (1) by interchanging p and q. But since only substitution (and not interchange) of letters has been postulated, the interchange had to be built up from substitutions.

If in (3) r is replaced by q one obtains

(3') (q implies (p implies q)) implies ((q implies p) implies (q implies q)).

(1*) and (3') yield by Modus Ponens

(4) (q implies p) implies (q implies q).

If in (4) p is replaced by the expression (p implies q) the result is

(4') (q implies (p implies q)) implies (q implies q),

which thus is an admitted formula. By Modus Ponens, (1*) and (4')
yield an important law of logic:

q implies q   (or p implies p).

The totality of all propositions thus derivable from the axioms is
called the *calculus of propositions*.

It is easy to show that Axioms (1), (2), (3) and all the consequences derived from them via (I) and (II) are tautologies. What
Post proved was that, conversely, each tautology is either an axiom
or derivable from the axioms via (I) and (II), in other words, that the
calculus of propositions coincides with the totality of all tautologies —
a fact also expressed by saying that Frege's system of axioms of the
calculus of propositions is *complete*.[6]

5.

Before listing a final group of Wittgenstein's theses I wish to discuss
an extremely important extension of the calculus of propositions
mentioned in the *Tractatus* that will often be referred to in subsequent
chapters. An inductive way to this extension is the observation of a
parallelism between certain propositions, e.g.,

'this crow is black', ... and the like.

Or with regard to a population P consisting of Swedes and Italians:

'the 1st Swede is taller than the 1st Italian',

'the 1st Swede is taller than the 2nd Italian',...

'the 2nd Swede is taller than the 1st Italian', . . .

'the 5th Italian is taller than the 3rd Swede', . . .

Frege was the first to distill, from series of this kind, *schemes of propositions* such as

'x is black', 'b is taller than c', and the like.

The letters x, b, c, . . . are referred to as the *variables* in the respective schemes. Such a scheme is not itself a proposition nor either true or false. It can, however, be transformed into a proposition by replacing all of its variables by names of individuals belonging to certain class agreed upon at the outset, which Edward V. Huntington called the *universe of discourse*, e.g. the class of all crows or the class of all visible objects or the population P.

Russell called such schemes *propositional functions* — clearly motivated by the following idea. Consider computing, for a given positive number x, the exponent to which 10 must be raised in order to yield x: one associates to x a number called the logarithm of x or, briefly, log x. For example, log 100 = 2 and the logarithm of 1/10 is -1 because $10^{-1} = 1/10$. This association or, more precisely, the resulting class of all pairs (x, log x), including e.g. (100, 2) and (1/10, -1), is called a *function*. In each pair belonging to the logarithmic function the first member, which is positive, is referred to as an *argument* of the function, the second is called the corresponding *value* of the function; e.g. the value of the logarithmic function for the argument 100 is 2. Since all values are *real* numbers (positive, negative, or 0, which is the value for the argument 1, since $10^{0} = 1$) the function is called a *real* function.

In the same way, the scheme 'x is black' associates to each name (as an 'argument') a proposition (as a 'value'), whence it is called a *propositional function*. Its values for the arguments 'that crow' and 'this snowball' are a true and a false proposition, respectively.

In addition to replacing variables by names of individuals, there is a second way to transform propositional functions into propositions. One may prefix to the propositional function a *quantifier*, i.e. the words *'for all'* or *'for some'* followed by the variable of the function, e.g.

'for all x, x is black', 'for some x, x is black'.

Instead of the latter proposition, one often writes 'there exists an x such that x is black' and refers to it as an *existential proposition*. Of course, a quantified proposition has a definite meaning only relative to a definite universe of discourse, population P in the second example. With regard to such a universe, quantified propositions can be formulated even without any use of variables, by saying, e.g.

some Swedes are taller than some Italians;

all Italians are taller than some Swedes;

and the like.

The restricted calculus of propositional functions deals with compounds of propositions and propositional functions by the logical connectives 'and', 'or', 'not', . . . , which may include the quantifiers, 'all' and 'some', with regard to individuals belonging to the universe of discourse.

Frege and, later, Russell developed systems of axioms for the restricted calculus of propositions.[7] Just as the (true or false) propositions about the blackness of crows, pieces of coal, snowballs, etc. are synthesized in the scheme 'x is black' one might synthesize the (true or false) propositions about various properties of a crow such as

>'this crow is black', 'this crow is feathered', 'this crow
>is white', . . .

in a scheme 'this crow is f'. Actually, however, one synthesizes the propositions about the various properties of an individual c by writing 'f c' rather than 'c is f.' Here, f is called a *property variable*, in contrast to the *individual variable*, x, in 'x is black.'

Even schemes of propositions, such as 'x is black', 'x is white', . . . are synthesized in the scheme 'f x', which stands for 'x has the property f', here, x may be replaced by 'this crow' or 'this snowball', and f by 'is black' or 'is white', in any combination, in order to obtain (true or false) propositions. Similarly one can construct schemes of propositions about relations between individuals a and b, such as F(a,b) and schemes of such schemes, such as F(x,y).

Two of the axioms of the calculus of propositional functions read

>(for all x, f x) implies f a,
>(for all x,y, f(x,y)) implies f(a,b).

The former is mentioned in the *Tractatus*, the latter sentence means: If, with respect to a certain universe of discourse, all pairs x,y are in a certain relation f, then the pair a,b is in the relation f, where a and b are individuals belonging to that universe. In IX,3 this axiom will be used several times in the proof of a simple geometric theorem, which could not possibly be inferred on the mere basis of the calculus of propositions.

The restricted calculus of propositional functions can be greatly extended by admitting quantifications applied to classes, propositional functions, etc., as in the following examples:

>in some classes of birds all birds are black;
>in all classes of birds some birds are black;
>in all populations some Swedes are taller than some Italians;

there are relations such that all Italians are in the relation to
some Swedes;
and the like.

Propositions of this kind are the object of the (unrestricted) *calculus
of propositions*, which likewise has been axiomatized.

6.

A third group of theses expressed in the *Tractatus* was debated in
the Circle during the years I attended the meetings: Wittgenstein's
treatment of numbers, of infinity and of probability; his views on
atomic propositions; his theory of language as a picture of the world
such that relations between the names in a proposition reflect relations between objects; and especially his claim that various aspects
of language, in particular the relations between language and the
world *evince themselves* (*zeigen sich*) but cannot be expressed in
the language. Some of these debates will be reported in subsequent
chapters.

One of Wittgenstein's ideas of the greatest consequence was his
claim (Tr. 5.54) that the truth or falsity of a compound proposition p
depends exclusively on the truth and falsity of its components and
not on their meaning or their content. This is certainly the case for
compounds by the logical connectives as studied in the calculus of
propositions, such as 'p implies q', 'p and (q or r)' and the like (see
VIII,3). But it is by no means clear for propositions such as 'I believe
that all men are mortal', which is true or false regardless of the truth
or falsity of the proposition 'all men are mortal.' But by a reinterpreta-
tion of propositions about belief (Tr. 5.542 sq.), Wittgenstein tried to
uphold his thesis, which greatly influenced Russell's writings.

7.

Summarizing what in the *Tractatus* immensely appealed to the members of the early Circle one must mention, first of all, the idea that whatever can, in a strict sense, be said about the world consists of (nontautological and noncontradictory) compounds of elementary or atomic propositions and that the only other propositions that can be asserted are the tautologies in the sense defined in this chapter, the latter saying nothing about the world. The attraction was enhanced by the statement (Tr. 2.223) that the truth or falsity of a picture can be ascertained only by comparing it with reality — a method that the Circle applied to nontautological propositions. In contrast, tautological propositions (which make up logic and have no content) are recognized as true by their mere form.

The Circle linked these views with classical philosophy by identifying the elementary propositions and their nontautological and noncontradictory compounds with *synthetic*, and the tautologies with *analytic*, propositions in the traditional sense. The former are clearly empirical; the latter are based on the grammar of the logical particles. There was no room for synthetic assertions a *priori*.[8]

This was approximately the (all too simple) view of the Circle after the first discussion of the *Tractatus*.

Notes

[1]Here and in the sequel, 'Tr.' stands for *Tractatus*. The numbers of the propositions in Tr. follow the system of numbering borrowed from Peano, which Whitehead and Russell used in the *Principia Mathematica*. But in contrast to books on mathematics, in which the main purpose of such numbers is to serve as short references to theorems in proofs of subsequent theorems, there are hardly any cross references or, for that matter, proofs in Tr. A 4-digit number in Tr. rather indicates that the proposition is a

comment on a comment on a comment on one of the seven propositions with a 1-digit number, e.g. Tr. 3.142 is a comment on Tr. 3.14, which is a comment on Tr. 3.1, which is a comment on Tr. 3. Similarly, Tr. 4.003 is a comment of 'minor logical weight' (less than that of, say, Tr. 4.01) on Tr. 4.

[2]Mauthner has indeed remained little known and badly underrated. In J. Passmore's in many ways excellent book *A Hundred Years of Philosophy* (2nd edition), he is not even mentioned, Weiler's book (see III,2) will probably make him better known in the English-speaking world.

[3]Here, 'or' is the so-called *inclusive* 'or', often called 'and/or' in legal language; that is to say, 'p or q'; means that *at least one* of the propositions p and q is true. The Romans expressed this usage by the particle '*vel*' and wrote '*aut* p *aut* q' to indicate that *exactly* one of the propositions p and q is true. The latter is parallel to the use of 'or', 'ou', and 'oder' in English, French, and German, where however, no sharp distinction between the two kinds of disjunction exists.

[4]*American Journal of Mathematics* 43, 1921.

[5]Lukasiewicz, Jan and Tarski, Alfred, 'Untersuchungen über den Aussagenkalkül', *Comptes Rendus des Séances de la Société des Lettres de Varsovie, Classe III*, v. 23 (1930) pp. 30-50.

[6]Moreover, Post's thesis contained the discovery of n-valued logics in which the propositions are divided into n classes of logical significance, not only into the two classes of true and false propositions — a discovery made simultaneously with and independently of Post by Lukasiewicz.

[7]Russell develops a cumbersome and unnecessarily complicated symbolism. He refers to the property of blackness or the predicate 'is black' by writing 'x is black.' More generally, for 'is f' or 'the predicate f belongs to' he writes 'f x' though the letter x has no meaning whatsoever and might as well be replaced by y or any other letter or be omitted. Occasions arise, however, where Russell must distinguish the predicate f from the propositional scheme f x and in such cases he uses for f the symbol f x, which still includes the superfluous letter x. He uses 'f' indiscriminately for predicates as well as relations instead of indicating them by symbols such as, f and F or f^1 and f^2, respectively. Yet, on occasion, Russell wants to indicate that 'f' at that place stands for a predicate. To this end, he introduced for the predicate (which he might simply denote by the letters f or f^1) the grotesque symbol f!x. This carrying along the

ubiquitous letter x in logical formulas, even where it means nothing and might be replaced by any other letter or simply omitted, is of course a heritage from the mathematicians, who since Descartes have indulged in a symbolism that I have once described as x-omantic.

[8] A priori means before, or at least independent of, experience. Kant had embarked on his critique of pure reason mainly in order to explain how synthetic (i.e., essentially, nontautological) propositions a priori were possible. Propositions that Kant regarded as synthetic and a priori included in particular all statements of Euclid's geometry. In the Circle, no one except the phenomenologist Kaufmann believed in the existence of synthetic a priori propositions.

IX. ON THE COMMUNICATION OF METAPHYSICAL IDEAS. WITTGENSTEIN'S ONTOLOGY

1.

In 1925, when I first heard about the *Tractatus* — I have forgotten from whom — I was busy finishing a paper 'Principles of a General Theory of Curves,' and this left me little time for anything else. Nevertheless I started reading the book. But I stopped after the first few pages. If this did not speak well for my instinct in matters philosophical, there certainly were extenuating circumstances even apart from the lack of leisure.

No one suggested to me in 1925 what some commentators recommend today to those who encounter the *Tractatus* for the first time: begin in the middle of the book. So I began at the beginning, particularly since the rigid numbering of all sentences in the *Tractatus* suggested a tighter cohesion of the propositions than the book actually affords.

I was greatly impressed by the Preface and especially by the beautiful remark that *what can be said at all can be said clearly*, followed by the even more important maxim, repeated in the last sentence of the *Tractatus*, that *whereof one cannot speak thereof one must be silent*. But when I started reading the first section of the book, sometimes referred to as Wittgenstein's *ontology*, I did not find the clarifying or even clear ideas I had expected. Some passages in fact appeared to me to be of typically metaphysical obscurity. Two years later, I learned through the Circle that the middle section of the *Tractatus* certainly would have fulfilled my expectation. But the ontology seemed to fall so far short of this that I did not extend my first reading beyond it.

Nor was my experience by any means unique. Before I entered the Circle, Hahn gave me a very brief synopsis of the latest discussions and asked me whether I had read the *Tractatus*. When told about my abortive attempt he said, "I must confess that after the first glance at the beginning of the *Tractatus* I did not think that the book was to be taken seriously. It was only after hearing Reidemeister's comprehensive report and carefully reading the *entire* work that I came to appreciate it as probably the most important philosophical publication after Russell's writings. The opening section belongs largely to that which according to Wittgenstein cannot be said."

Wittgenstein indeed compared the statements in the *Tractatus* to "a ladder that one has to throw away after having climbed up it" (Tr. 6.54) — a simile that Mauthner had used verbatim in describing his own work more than a decade earlier. But, as the *Tractatus* demonstrated, what cannot be *said* in a strict sense can still be *talked about* — talked about or discussed or whatever term one would use for not being silent about the ineffable. This has been frequently pointed out, first in Russell's Introduction to the book itself.

Equally important, however, is the fact that the converse of Wittgenstein's remark is *not* valid; in other words, *it is not true that what can only be talked about can only be talked about unclearly*; and this fact does not seem to have been sufficiently emphasized.

So, after my conversation with Hahn, I tried to pin down the reasons for my difficulties with the beginning of the *Tractatus* by a careful analysis of the passages I found unclear. My opinion in this matter has not changed during the almost fifty years that have passed since then. On the contrary, I now feel that the scope of that analysis transcends Wittgenstein's ontology and embraces meta-physical ideas in general. In this chapter therefore I shall attempt to summarize my point of view — the point of view of a reader, especially a mathematically trained reader, expecting a clear com-munication of metaphysical ideas.

2.

Every oral or written communication of ideas is conveyed in sen-tences, which are in turn made up of words.

A method by which some naive minds, even today, seek to ascertain the validity of communication is to define all words and to prove all sentences. But as Pascal, wrote in his opuscule *De l'esprit géométrique* more than three centuries ago: *"Cette méthode serait belle . . . mais elle est absolument impossible."* This is because words are defined in terms of other words, and sentences proved from other sentences. A *regressus ad infinitum* being impossible, the process of defining and proving must necessarily terminate at some stage and at that very point the communication starts. It starts with *un*proven sentences made up of *un*defined words.

The undefined words and unproven sentences in *one* commu-nication may of course be defined and proven in *another* one. The analytic geometry of the plane, for instance, which deals with pairs of numbers, includes unproven sentences about numbers; and num-bers remain undefined. Explicit or implicit definitions of numbers and proofs of those sentences about them can, however, be found in arithmetical theories, which start with unproven sentences consisting of undefined words of their own.

While this situation is universally recognized in geometry, it is by no means generally understood or admitted, let alone applied, by philosophers. Yet it is a general principle, valid for philosophical as well as geometric, scientific or, for that matter, practical discussions that *every communication must start with unproven sentences consisting of undefined words.*

An all-time ideal for the communication of ideas was created in Euclid's *Elements*, one of the greatest books ever written. The ideal, which the book itself does not fully attain, consists in the presentation of an initial list of sentences, sometimes referred to as *postulates* or *axioms*, which logically imply the remainder of the communication. All that is needed for the comprehension of the communication is the understanding of those initial sentences and the operative knowledge of logic.

Clearly, the basic sentences include all the extralogical words that occur in the entire communication except possibly some that are introduced by definitions in terms of words included in the basic sentences. Most of these defined words serve as abbreviations of frequently occurring expressions; and all of them can always be replaced by the expressions for which they stand. Extralogical words that are not present in the basic sentences or replaceable by words that are present cannot possibly occur in sentences logically derived from the basic ones.

The insight into this situation, first achieved in Pasch's geometric work (see IV,2), has resulted in the following modern view. A deductive theory should start with two lists of its *'primitive elements'*: a list of all *undefined words* and a list of all *unproven sentences*. The development of the theory is the deduction of consequences of those postulates.

For the benefit of nonmathematical readers, Section 3 will be devoted to the elements of such a deductive theory — a very simple theory, but developed with that absolute logical rigor that was only introduced into theoretical thinking about two thousand years after Euclid.

3.

The following theory deals with a particular aspects of the geometry of the Euclidean plane and is called the *affine plane geometry*. In it, one ignores angles, perpendicularity, and congruence, and one centers the attention entirely on points, lines, and their incidence. An approximate empirical model is supplied by a blackboard and certain entities on it: chalk dots and lines drawn by means of a straight edge and the relation described by saying 'a point is on a line' or, synonymously, 'a line is on a point.' These physical entities and observations concerning them suggest the choice of the undefined terms and of the assumptions of the theory.

UNDEFINED TERMS: *point, line, on.*

The second list includes three propositions asserting what is usually expressed by saying 1) that any two points are joined by exactly one line, 2) that there exist three points that are not collinear, and 3) that Euclid's famous Parallel Postulate holds. In the late eighteenth century version of J. Playfair, this postulate lays down that for each line, through each point not on it, there is exactly one parallel line. These three unproven propositions must, however, be formulated in such a way as to include only logical particles (the numerals 'one', 'two', 'three' being regarded as abbreviations for expressions made up of logical terms), and the three undefined terms. In particular, parallelism will be introduced by the following

DEFINITION: Two lines are *parallel* if and only if no point is on both lines.

In these terms, the short list of assumptions reads as follows:

UNPROVEN PROPOSITIONS (POSTULATES):

I. For any two points there is
   a) at least one,
   b) at most one
   line such that both points are on the line.

II. There are three points that are not on any one and the same line.

III. For any line, any point not on it is on
   a) at least one,
   b) at most one
   parallel line.

If one adjoins a certain fourth primitive proposition to the theory, then one can deduce a considerable part of plane geometry (including e.g. well-known theorems about triangles). But more important for philosophers than these technicalities should be the realization that even the simple Postulates I, II, III, yield non-trivial consequences. For instance, while the postulates themselves only state that there are three points, they imply that there exists at least one other point[1] and that there are several lines. I shall formulate this result and its proof as a simple example of a truly rigorous deduction.

THEOREM. Under Assumptions I, II, III, there are at least 4 points and 6 lines. The 4 vertices and 6 edges of a tetrahedron constitute a minimal system satisfying those postulates.

PROOF. By Postulate II, there are 3 points, say P, Q, R, that are not on any one and the same line. By I.a, P and Q are on a line, say *l*. Since P, Q, R are not on a line, R is not on *l*. Hence by III.a, R is on a line *l'* that is parallel to *l*.

By I.a, P and R are on a line, say *m*. Hence Q is not on *m*. By III.a, Q is on a line, say *m'* that is parallel to *m*. Being on *m*, the point P is not on the parallel line *m'*. But P is on *l*. Hence *l* and *m'* are different lines. The lines *m'* and *l'* are not parallel. For if they were, then Q would be on two different lines, *m'* and *l*, both parallel to *l'* (see Figure IX.1), whereas, by III.b, Q is on only one line parallel to *l'*.

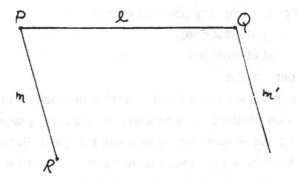

Figure IX.1

Since *l'* and *m'* are not parallel there is a point, say S, that is on both *l'* and *m'*. Since P and Q are on *l* and R is on *m* while S is on *l'* (parallel to *l*) and on *m'* (*parallel to m*) it follows that S is different from P, Q, and R. Thus S is a fourth point (See Figure IX.2).

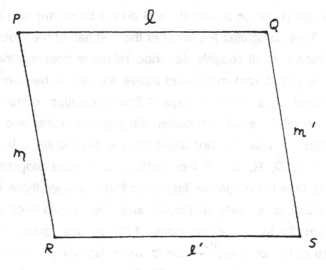

Figure IX.2

By l, Q and R are on a line, say n; and P and S are on a line, say n'. That there are at least 6 lines will now be established by proving that the lines

l, l', m, m', n, n'

are pairwise different. The lines l and l' as well as m and m' are parallel and hence certainly different. Furthermore, it is clear that

(1) P is on l, m, n' but not on l' or m';

(2) Q is on l, m', n but not on l' or m;

(3) R is on m, l', n but not on m' or l;

(4) S is on l', m', n' but not on l or m.

From (1) it follows that, in addition to the pairs (l,l') (m,m'), and (l,m'), which have already been discussed, the following pairs are different; (m,l'), (n',l'), and (n',m'). (2) adds the pairs (l',m), (m',l'), (n,l') and (n,m'); (3) the pairs (n,l) and (n,m'); (4) the pairs (n',l) and (n',m). It remains to be shown that n and n' are different. But P is on *n'* while Q and R are on *n*, and P, Q, R do not lie on any one and the same line. This completes the proof of the first half of the theorem.

Postulates I, II, III roughly describe relations between the chalk entities on a blackboard mentioned above as well as between pencil dots and lines on a sheet of paper.[2] But in addition, there exist of course many other relations between the physical points and lines in those models — relations *not* expressed in Postulates I, II, III. For example, if P, Q, R, S are the vertices of a parallelogram on a blackboard, then the diagonals intersect; that is to say, there is a fifth point, T, which is on both diagonals: and T is on two lines each of which is parallel to two of the sides of the parallelogram. Hence a blackboard does not contain 4 points and 6 (straight) lines which, by themselves, satisfy Postulates I, II, III. But as can readily be verified, these assumptions *are* satisfied by the 4 vertices and 6 edges of a tetrahedron (if one ignores the points on the edges other than the vertices). Opposite edges are parallel in the sense of the definition of parallelism, since none of the 4 vertices lies on two opposite edges (see Figure IX.3). They thus represent a *miniature affine plane*, more precisely, a minimal system satisfying Postulates I, II, III, as the second half of the theorem claims.[3] This completes the proof of the theorem.

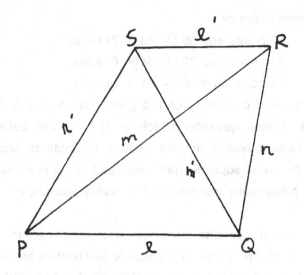

Figure IX.3

It can furthermore be rigorously proved that the next larger systems satisfying I, II, III consist of

9 points and 12 (= 9 + 3) lines,
16 points and 20 (= 16 + 4) lines,
25 points and 30 (= 25 + 5) lines.

It follows that there are no systems satisfying I, II, III that include, say, exactly 5 or 12 or 17 or 24 points or exactly 16 or 29 lines.

A nonmathematician inspecting the preceding list of systems might suspect that these last results have been obtained without great difficulties since he himself might well venture a conjecture as to the next larger system: to wit, that it consists of 36 points and 42 (= 36 + 6) lines. Actually, however, these results are not at all easily obtainable; and the nonmathematician's conjecture is false: no system of 36 points can satisfy I, II, III. The next larger 'finite affine planes' consist rather of

49 points and 56 (= 49 + 7) lines,

64 points and 72 (= 64 + 8) lines,

81 points and 90 (= 81 + 9) lines.

Whether a system consisting of 100 points can satisfy I, II, III is an exceedingly difficult question, which in 1971 is still baffling some excellent mathematicians despite serious attempts to answer it by the use of the most sophisticated methods.[4] But it is known that the next larger affine plane consists of 121 points and 132 (= 121 + 11) lines.

A blackboard contains no finite systems of points and lines satisfying I, II, III. What then is a possible justification for the study of those phantastic creations of the mathematical mind? First of all, the theory of finite planes is of great beauty comparable to that of classical music, (and nobody asks about a justification for music). The theory, moreover, represents a profound and thought-provoking synthesis of geometry, combinatorics, and number theory. Besides, finite planes are applicable, especially to statistics. More specifically, they are useful in the design (i.e. the planning) of experiments. When several chemicals are to be tested as to their effects on the soil the experimenters divide a large field into parcels, each designated to test one of the substances. The most efficient and significant divisions of the field are obtained on the basis of results about finite planes. Thus the U.S. Department of Agriculture makes practical use of finite planes, and even a very earthy use, namely in testing fertilizers.

Thus this section has demonstrated that a geometer starting from a few utterly simple assumptions and proceeding by rigorous logic, may arrive at a variety of consequences — some expected, others unexpected — and at extremely difficult, (though perfectly clear) problems.

4.

Euclid's *Elements*, though regarded as a model of logical rigor for two thousand years, actually did not fully attain the ideal that it itself created. Two serious shortcomings have become apparent.

In the first place, some of Euclid's proofs make essential use of tacit assumptions, which are missing from his list of postulates and cannot be derived from those that are listed. These gaps were discovered by Pasch and filled by Pasch, the school of Peano, and Hilbert (see IV,2). One of the most striking of these shortcomings of the *Elements* concerns the *arrangement* of points on a line, about which the book formulates no assumptions at all, even though many proofs are based on the existence and properties of such arrangments. Pasch was the first to explicitly formulate simple sentences from which he deduced that the points on each line can be *ordered* in the same way as the points on a scale. To this end, he introduced a new undefined term, *between*. It describes a ternary relation for points, subject to a few simple postulates including the following:

If Q is between P and R, then 1) Q is between R and P; 2) R is not between P and Q; 3) if, in addition, R is between P and S, then Q is between P and S, and R is between Q and S; 4) if Q is between P and R, and R is between Q and S, then Q as well as R is between P and S.

In the second place, not only do the *Elements* lack a list of undefined terms, but Euclid does not seem to have realized the inevitability of undefined terms, or to have seen clearly what undefined terms he was actually using. Instead, the *Elements* began with definitions. The first one is the famous sentence '*A point is that which has no parts*'. The term *part* is (and remains in the *Elements*) undefined and all that is assumed about it, is that the whole is greater than the part. So none of the numerous theorems or proofs makes use or could make use of the definition of point. Actually, *point* is thus a primitive term of Euclid's theory, and so is essentially the term *line*.

The idea of strictly deductive theories (derived from unproven sentences that include, besides logical particles, only undefined words) was widely and successfully propagated by Hilbert under the name of the *axiomatic method*. It has been applied to logic and algebra, to parts of physics and chemistry, to genetics and the theory of family relations, to economic theory and sociology. Philosophers must carefully avoid two errors related to this idea.

The first misconception, which has evolved over the centuries and is connected with Euclid's Elements, is the idea that the undefined words and unproven assumptions are so *clear* that they are *not in need*, and so *simple* that they are *not capable*, of a definition or a proof, respectively. Actually, geometers questioned Euclid's Parallel Postulate, III, for two thousand years, until at the beginning of the nineteenth century some ingenious mathematicians boldly based a 'non-Euclidean' geometry on the negation of III.b (see IX,3, endnote) and later, another one on the negation of III.a. In other fields, the axiomatic method has resulted in the development of important theories from primitive elements some of which are certainly not simple while the clarity of some others is controversial. There are fields in which clear and simple primitive elements do not seem to exist. Yet even there Pascal's Principle applies: defining and proving must be terminated at some stage. Clearly, *this principle applies to every communication of ideas* and, therefore, cannot be ignored even by those who intend to communicate metaphysical ideas. Hence the belief in the irrelevance of the axiomatic method for philosophy is another misapprehension of which philosophers must beware.

Section 6 will elaborate from this point of view on the difference between philosophical and geometric theories. But first, Section 5, will be devoted to Spinoza's attempt to apply Euclid's original procedure to philosophy.

5.

Since Euclid's work, for all its greatness and historical importance, fails to satisfy a modern logician or geometer, it is not surprising that Spinoza's *Ethics* (*Ethica ordine geometrico demonstrata*), which is loosely modelled after the *Elements*, is even less satisfactory. The *Ethics*, too, begins with definitions — Part I with eight definitions, of the terms *self-caused, finite in its kind, substance, attribute*, etc. Hence the modern reader is forced to distill the undefined terms from those definitions. In this way, he arrives at the terms *thing, essence of, existence of, nature of, limited by, to exist in, conceived through*, etc. The 7 axioms followed by the definitions introduce further undefined terms, e.g. *cause, effect, follow, knowledge of, involve*, etc. The remainder of Part I is made up of thirty-six propositions including the statement that *each substance is necessarily infinite*. Each of these propositions is followed by what Spinoza (in most cases quite mistakenly) considers to be a proof.

Axioms I and II state that what exists, exists in itself or in something else; and that everything is conceived through itself or through something else. If, instead, Spinoza had merely postulated that *what exists exists in something* and that *everything is conceived through something*, then these weaker assumptions would imply his Axioms I and II by virtue of the law of the excluded middle. This would be an extremely simple proof, but, in contrast to Spinoza's, a valid one.

Axiom III includes the terms *cause, effect,* and *follow.* It states that from a definite cause an effect necessarily follows, whereas without a cause no effect can possibly follow. Since the terms *cause* and *effect* do not occur in any other axioms of Spinoza's it is clear that little can be proved about them in the strict sense.

A logician who wanted to do for Spinoza's *Ethics* what Pasch did for Euclid's *Elements* would face a more difficult and less rewarding task than Pasch did. Some parts of the *Elements* could be retained without modification; other parts required only an elaboration that took the newly formulated assumptions concerning order into account; few parts besides the theory of order had to be built from scratch. The *Ethics*, on the other hand, would have to be completely rebuilt, beginning with a very extended list of primitive terms and a greatly enlarged list of postulates. And since Spinoza's concepts are hardly interrelated in a rigorous sense it is doubtful whether a modern treatment of the contents of that book would yield anything beyond a serial or ramified arrangement of unproven propositions (including the majority of those that Spinoza called theorems). Thus, after a thorough analysis of the terms and propositions, one would be left with many assumptions yielding few, if any, logical consequences. But a mathematician or logician considers his activity rewarding only if, on the contrary, he has a chance to derive many consequences from few assumptions.

6.

A comparison of the *Ethics* with the *Elements* or, better, with a rigorously constructed modern geometric theory such as the simple example in IX,3 brings the differences between metaphysical and geometric material into full relief.

The first dissimilarity, already mentioned in the preceding section, lies in the fact that metaphysical concepts are not as strictly interrelated as are, e.g. the ideas of points and lines in the postulates of affine geometry; nor are the various metaphysical sentences as closely interconnected as are the geometric postulates. Yet such strict interrelations and close interconnections are by no means confined to geometry. They exist, to a greater or lesser extent, in all the domains in which axiomatic theories have been developed, from logic to sociology (see IX,4); and in principle, it is of course conceivable that metaphysical concepts might also be interconnected in sentences that yield nontrivial stricty logical inferences. But it is hard to see how this might be done for concepts such as *essence* and *existence, substance* and *attribute, the absolute* and the like; and, at any rate, it has, to the best of my knowledge, never been done so far.[5]

The second difference concerns the understanding of the meaning of the extralogical terms, many of which are borrowed from the everyday language. Strictly speaking, a consensus on this point is irrelevant since a deductive theory is entirely based on its postulates; and neither empirical facts nor intuitive pictures associated with the sentences, nor any connotations of the terms must be utilized in its development or its comprehension. Yet it is a help both in developing and in understanding a simple geometric theory that everyone associates more or less the same ideas with the words *point* and *line* (as is evidenced by the absence of misunderstanding about these top-ics). In the transition from simpler to more complicated theories, the assumptions change and, consequently, so do the implicitly defined meanings of the terms. They change gradually, but eventually they change beyond recognition. For example, in the axiomatics of the so-called complex plane, the points are assumed to behave like pairs of complex numbers and there are lines that are perpendicular to themselves![6] And yet in the field of geometry, each such 'advanced' theory is connected with the universally understood sim-plest developments by a chain of intermediate theories. But with terms such as *substance, essence*, and the like, consensus about their meaning is lacking from the very outset. Two persons reading in Spinoza's *Ethics* that each substance is necessarily infinite, may connect totally different ideas, (or no ideas at all) with that proposi-tion.

These differences between metaphysical and geometrical ma-terial have various consequences.

The paucity of logical interrelations and interconnections reduces metaphysical theories, as already observed, essentially to serial or ramified arrays of propositions with few, if any, proofs or nontrivial consequences — a situation similar to that occurring in descriptive natural science or in some religious creeds. For example, the Muslim creed is the simple conjuction of seven theses: I believe in Allah *and* in his angels *and* in his prophets *and* in his books, etc. As an extreme case, one may consider total unrelatedness. Take, say, three sentences p_1, p_2, p_3, such as the nonphilosophical proposi-tions: 'gold is a metal', 'Paris is the capital of France', 'lions are mammals', which jointly include 6 or 7 extralogical terms, each of them occurring in only one of the sentences. Like every set of propositions, this triple of sentences has consequences even apart from tautologies such as 'p_1 or not p_1'. For example, the consequences include 'p_1 and p_2', 'p_1 or p_2 or p_3', 'p_2 or not p_3' and, more generally, all those compounds with components p_1, p_2, p_3, which are true, if (though not necessarily only if) all components are true. I shall call these *the trivial consequences* of the three propositions. They are consequences that philosophers would hardly care to list as such. But the three mentioned sentences have no other consequences.

An array of unrelated sentences may be unrewarding to the logician. But it is impeccable as long as the writer refrains from claiming that logical relations exist. Unfortunately, however, just such pretensions are not uncommon in metaphysical writings. They are demonstrated by unjustified, thoughtless use of particles such as 'hence', 'therefore', 'consequently', and the like. This is a practice which even the most tolerant logician must reject.

The lack of consensus concerning the meaning of some metaphysical terms results in a groping after definitions. So we get sentences expressing one word in terms of others that have been introduced in sentences preceding the one under consideration. This procedure is reflected in frequent changes of some of those preceding sentences — changes in content, not only in form! — and the retraction of the original formulations. This, too is a familiar situation in descriptive natural science, where new discoveries necessitate perpetual restatements, just as *all swans are white* had to be restated after the discovery of Australia. A logician sympathetic with informality in presenting logically unconnected metaphysical sentences may condone this practice. For while it has been traditional since Euclid to place at the beginning of a theory, sentences which, as we know since Pasch, serve as implicit definitions of the extralogical terms, it is not logically necessary to do so; and consistent systems of reformulation are not illegitimate.

While the majority of the words that occur in metaphysical studies are borrowed from everyday language there are also neologisms. For the purpose of the communication of ideas the latter must be explained in terms of the former — perhaps gradually and gropingly, but somehow. While the overindulgence of some metaphysicians in coining new terms is irritating to the logician and often appears to him a symptom of charlatanism, a much more insidious practice is the deviant use, so common in metaphysics, of words of everyday language. Yet it is just such procedures that often produce metaphysical problems and studies. When stripped of such artificially created difficulties most metaphysical studies can be reduced to arrays consisting of a few sentences. For each of the theories known today, the sentences of the corresponding arrays lack non-trivial logical relationships.

The preceding critical discussion has been strictly confined to a single branch of what is traditionally called philosophy, namely metaphysics. Philosophy in the Mauthner-Schlick-Wittgenstein sense of a clarifying activity, however, is not an object of the preceding criticism. It is true that a philosophical discussion of this kind too starts with unproven sentences made up of undefined words. But the most important of those sentences are methodological suggestions, advice, warnings and the like. And the terms that they include, such as 'term', 'sentence' and the like, are (for the purpose of the discussion!) universally understood. Lists of the primitive elements of those clarifying discussions can be, and have been, provided.

To summarize: Metaphysicians who propose to communicate cognitive ideas to others in prose (rather than in the form of poetry, or drama, music, or dance) must willy nilly start with unproven sentences made up of undefined words. *How else could they start*? No rigmarole can obscure Pascal's Principle as applied to metaphysics. The primitive elements of *known* metaphysical theories, however, have never been satisfactorily listed. On the other hand, I consider it by no means impossible, that in the future some metaphysical theories will be developed on a logically adequate basis.

In Section 7, I propose to scrutinize the ontology in the *Tractatus* soberly from the point of view that has just been developed.

7.

The ontology (as some have called it, though not the author) of the *Tractatus* is presented in about eighty propositions (labelled Tr. 1 - 2.225). They are free of neologisms and of that inane talk that characterizes so many metaphysical writings. In contrast to Spinoza's *Ethics* and a few other metaphysical treatises, the ontology is also free of pseudoproofs; and in a few cases where statements are derivable from others, the simple valid proofs are not presented. There is hardly even any implied claim that sentences are logically interrelated.[7] The propositions are rather presented in that simple serial (or, by virtue of the numbering, ramified) arrangement that would result from a thorough analysis of the *Ethics* (see IX,5). The sentences in the ontology, however, cannot be said to be the result of a thorough analysis.

Distilling the undefined extralogical terms from the first nine propositions (Tr. 1 - 2.01) one finds

the nouns: world, fact, thing (synonymous with object), logical space, compound, existence; *Sachverhalte*;

the predicates: is the case, to break up into, can be.

According to Tr. 2, the terms 'fact' and 'what is the case' are synonyms. With this in mind, one can render the first four propositions (except for a side issue to be commented on below) in the simple sentence: *The world is the set of all facts.* Everything else in those four sentences is a play with the superfluous synonym 'what is the case' and an implicit description of the set concept (parallel to statements in book on set theory written at the turn of the century: *a set is determined by its elements* or *by what is, and what is not, an element of the set*).

The very next, fifth, sentence (Tr. 1.13) reads *"The facts in logical space are the world."* Whatever, the term 'logical space' (which recurs for the first time only four pages later) may mean, there clearly are only two possibilities: Either 1) *some but not all facts are in logical space*, in which case Tr. 1.13 is a qualification of the almost immediately preceding statement Tr. 1.1, according to which the world is the totality of *all* facts. In this case, it was superfluous and only confusing to mention Tr. 1.1 at all instead of saying right away that the world is the set of all those, and only those, facts that are in the logical space. Or 2) *all facts are in logical space*, in which case *just this* should be said without confusing the proposition by a superfluous reference to the world.

The sixth sentence (Tr. 1.2) states: *"The world breaks up (zerfällt) into facts."* Since the undefined term 'break up' does not seem to recur, one must appeal to its daily use according to which every set, in a way, breaks up into its elements. It thus appears that Tr. 1.2 does not add anything to the proposition that the world is the set of all facts. The next sentence: *"Any fact can be or not be, anything else remaining the same"*(Tr. 1.21),[8] seems to claim that facts are mutually independent in some unspecified sense (only vaguely hinted at by the use of the undefined term 'can'). But since, in *some* sense, facts are obviously interdependent, an author desirous to communicate ideas should have found it imperative to specify in *what* sense any fact 'can be or not be.'

The next sentence (Tr. 2), one of the seven principal propositions of the book, states: *"What is the case, the* (sic) *fact,*[9] *is the existence of Sachverhalte."* the German word '*Sachverhalt*' has no accurate counterpart in English. Possible translations include: *the way things stand, situation, circumstance*, or *state of affairs* (adopted in the second translation of the *Tractatus*). According to Tr. 2.01, *a state of affairs is a compound of things*. In Tr. 1.1 Wittgenstein has also pointed out — this is the side issue referred to above — that the world, the set of all facts, is not the set of all things. From what has been said so far, it follows that *the world is the set of all those compounds of things that constitute states of affairs.*

This concludes the first nine sentences. I will not discuss the next group of propositions which include the connotation-loaded terms 'essential', 'accidental', 'possibility', 'can be', etc. The presence of the word 'essential' seem to make it futile to comment from a logical point of view on statements such as *"It is essential to the* (sic) *thing that it can be the* (sic) *component of a state of affairs,"* (Tr. 2.011).

Tr. 2.02 states that *"the object is simple"* and Tr. 2.021 explains: *"The objects constitute the substance of the world; therefore, they cannot be composite."* On the basis of all propositions preceding Tr. 2.021, its second half is a *non sequitur*. The conclusion becomes valid only on the basis of the tacit assumption that *substance is simple* — a statement no more convincing, nor even more meaningful than Spinoza's assertion that each substance is necessarily infinite (see IX,5). Tr. 2.024 explains substance as *"what exists independently of the world."*

Therefore what we are told can be summarized as follows: The world is the set of certain compounds of things; the latter, being the substance of the world, exist independently of the world . . .

Notes

[1] A confusion of this assertion that certain points and lines exist with metaphysical statements about existence should be forestalled by the following explanation. *There is* (or *there exists*) a fourth point means: If any set of two kinds of entities, called points and lines, satisfies Postulates I, II, III, then this set includes, besides the 3 points mentioned in II, at least one more of the entities called points.

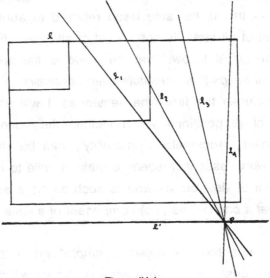

Figure IX.4

[2] Actually, assumption III.b does *not* hold in those models; e.g., it does not hold in the smallest of the rectangles in Figure IX.4 for the point P is on the lines l', l_1, l_2, l_3, l_4, and on countless other lines each of which is parallel to the line l in the sense of the definition; that is to say, no point of the sheet is both on that line and on l. On the larger sheet of the size of the middle rectangle, the extension of l meets those of l_1 l_2 while being parallel to l_3 and l_4. On the still larger sheet of the size of the largest of the

three rectangles, the extension of l meets even those of l_3 and l_4. But P is still on two unlabeled (and on countless undrawn) lines which are parallel to l. Yet it appears that a *sufficient* extension of *each* line (other than l') meets an extension of l. At any rate Euclid assumed that, in the *whole* plane, l meets *all* lines on P except one line l', which is *the* line on P parallel to l. Therefore, III is called Euclid's Parallel Postulate. At the beginning of the 19th century, the Hungarian Janos Bolyai and the Russian Nicolai I. Lobachevsky developed a *non-Euclidean* geometry (now also called *hyperbolic* geometry, (see IV,2) on the basis of the assumption that even in the whole plane P is on many lines parallel to l.

[3]That a 3-dimensional space includes 4 points and 6 lines satisfying Postulates I, II, III while the plane does not, in any way, indicate that these posulates describe space. Any axiomatic theory of space must, among other things, bring out 1) that besides points and lines there are planes; 2) that a line l and a point P not on l are on exactly one plane; 3) that on that plane P is on exactly one line parallel to l, while 4) P is on many lines l_1, l_2, . . . (said to be *skew* to l) which are not on that plane and yet have the property that no point is both on l and l_1, or on l and l_2, . . . What that fact shows is merely that the simple definition of parallel lines used in the affine geometry of the plane is inadequate in the geometry of the space. In considering the 4 vertices and 6 edges of a tetrahedron as the systems of points and lines of a miniature affine plane one ignores the 4 planes (the faces of the tetrahedron) and thereby makes it possible to call opposite edges parallel even though, in space, they are skew.

[4][The question of the existence of a finite affine plane of order 10, i.e., consisting of 100 points, was answered in the negative in 1989 by means of sophisticated computer methods. See C. W. H. Lam, L. H. Thiel, and S Swiercz, 'The non-existence of finite projective planes of order 10', *Canadian Journal of Mathematics*, XLI (1989) pp 117-1123, and C. W. H. Lam, 'The Search for a Finite Projective Plane of Order 10', *The American Mathematical Monthly*, v. 98, no.4 (April 1991) pp 305-318. It should be noted that the existence of a finite projective plane of a given order implies the existence of a finite affine plane of the same order and conversely. L. G.]

[5]Connections between the ideas of *souls* and *bodies* or the reconciliation or limitation of the ideas of *omniscience, omnipotence, omnipresence*, and similar faculties might offer a glimmer of hope for the development of logically correct, if simple, metaphysical theories, the former with a Western and an Indian variant. Of course, even if such a

theory were created some positivists would brand it as a sterile play with words. But this need not be a deterrent to anyone who aims at a theory with the aesthetic qualities that distinguish some geometric and algebraic developments, quite apart from the fact that geometers and algebraists have seen some of their ivory tower developments applied to reality — the theory of finite planes to testing fertilizers (see IX,3)! The difficulty lies in interrelating metaphysical ideas in nontrivial ways. If this could be achieved, chances are that in time someone would discover actual phenomena exhibiting those nontrivial relationships.

[6] In the analytic geometry of the complex plane, points are *defined* as pairs of complex numbers, just as, in the analytic geometry of the real plane, points are defined as pairs of real numbers. And, just as in the latter theory, lines with the slopes m and m' are said to be perpendicular if and only if $m \cdot m = -1$. Since $i \cdot i = -1$, each line of slope i is, therefore, perpendicular to itself.

[7] On the first pages, I find one correct 'because', one *non sequitur* introduced by 'therefore', and one questionable 'evidently'.

[8] This is my own translation of the not quite idiomatic German version of Tr. 1.21. When I began reading the *Tractatus* in 1925, Tr.1.21 was the first statement that struck me as quite dogmatic. It may be one of those that Russell had in mind when comparing some of the sentences in the *Tractatus* with a Czar's ukases — a comparison that, strangely enough, he made only 40 years after first seeing the book.

[9] Because of Wittgenstein's totally unsystematic use in the German original of singulars and plurals, definite and indefinite articles, etc. (some of which I point out by pedantic sics), it is not clear to me whether by a fact he meant the existence of only *one* state of affairs or possibly of *several* ones. Equally loose and unsystematic is the use in the *Tractatus* of the terms 'existing' and 'existence of.' Tr. 2.04 identifies the world with the set of all *existing* states of affairs after it has been identified with the set of all facts, and a fact has been said to be the *existence* of states of affairs. Some sentences are completely incomprehensible, e.g., "If the objects are given, then therewith *all* objects are also given" (Tr. 5.524). In the bilingual editions of the *Tractatus*, some of the shortcomings of the German original have been corrected in the English translation. In the next paragraph, I translate the German version of Tr. 2.011 *verbatim*.

X. WITTGENSTEIN, BROUWER, AND THE CIRCLE

1.

Schlick's personal relations with the author of the *Tractatus* began in 1927. From the outset Schlick was greatly impressed by Wittgenstein and deeply regretted his having given up work in philosophy. In this connection, I remember a party sometime after Schlick had returned from England.[1] The conversation turned to some philosopher or scientist who had stopped working and someone brought up Wittgenstein. "There is a great difference, though," said Schlick. "Mr. (whoever it was) ceased working because of fatigue, whereas Wittgenstein has given up work because of *ressentiment*." I remember Schlick using the French word, which incidentally has no precise German counterpart and, in a way, is slightly stronger than the English 'resentment.' ("Souvenir d'une injure, désir de s'en venger," explains Larousse and adds the example "conserver le vif ressentiment d'une offense.") At this point, the conversation unfortunately was interrupted by someone's joining the group and was not resumed despite the questions it raised in my mind. Then and later various signs seemed to me to point to Wittgenstein's particular resentment against mathematicians. Schlick's awareness of this fact may have been one of the reasons why, in 1927, he introduced Carnap, Feigl, and Waismann (who was by then definitely a philosopher rather than a mathematician) to Wittgenstein while keeping mathematicians away from him.

2.

In March of 1928, a few weeks after that conversation, it was announced that the Dutch mathematician L. E. J. Brouwer would give two lectures on philosophy of mathematics. These talks were to be the first in a series of guest lectures by foreign scientists — about three speakers each year. The series was financed by a few industrialists whom the physicist Felix Ehrenhaft had interested in this project. The lectures, to be held in the auditorium of one of the university's institutes of physics, were open to the public. As it turned out, all lectures in that series were well attended by students and faculty as well as by professional and business men interested in science.

Brouwer's two talks were entitled 'Mathematics, Science, and Language' and 'The Structure of the Continuum.' While none of the other members of the Circle knew Brouwer as a lecturer, I had often heard him speak in Amsterdam. So, thinking of the conversation with Schlick a few weeks before, I said to Waismann "Why don't you invite Wittgenstein to these lectures? Brouwer is a stimulating speaker and his topics may arouse Wittgenstein's interest." Waismann reflected for a split second and said "That is a very good idea. I will speak to Feigl. Perhaps we can induce Wittgenstein to attend." Two days before Brouwer's lecture, Waismann told me that Wittgenstein would be present.

3.

Hahn, who was to introduce Brouwer, was notified by Waismann or Feigl when Wittgenstein entered the auditorium. (Like myself, he had until then seen Wittgenstein only in a photograph, which incidentally proves that that latter had never attended a meeting of the Circle even in its first years.) From a distance I watched Hahn walking down the aisle to introduce himself and to welcome the guest. Wittgenstein thanked him with an abstract smile and eyes focused at infinity, and took a seat in the fifth row or so.

I have always tried to avoid making the acquaintance of some-one who appeared to be not interested in making mine and thus stayed far away. But I was curious to see how the guest would behave during the lecture which he attended, probably unbeknownst to him, at my suggestion. So I took a seat two rows behind him and well to his left. Motionless from beginning to end, Wittgenstein looked at the speaker with a slightly startled expression, at first, which later gave way to a faint smile of enjoyment.

4.

Brouwer's first lecture consisted of two parts. The first of these consisted of enigmatic remarks about 1) *mathematical contemplation* (mathematische Betrachtung) — a notion he used in the extremely wide sense of a certain voluntary activity justified by its appropriate-ness in achieving aims; 2) the way by which mathematical con-templation leads to what he called the *primordial phenomenon* (Urphänomen) *of the intellect*, namely, a sequence of arbitrary length (Presumably, Brouwer was thinking here of the contemplation of longer and longer sequences of objects such as rows of trees or of stones, on a prearithmetical level, counting being reserved for the next stage); 3) the substratum of all dyads (das Substrat aller Zweiheiten). This completely obscure idea is, according to Brouwer, obtained on higher levels of culture by abstraction developed from the mechanism of mathematical activity and constitutes what he called the *primordial intuition* (Urintuition) of all mathematics. The unfolding of that intuition yields, as Brouwer claimed, without in the least elaborating on this point, first the natural, then the real numbers, and finally the whole of pure mathematics. In *science*, the effectiveness of mathematical abstraction is due to the fact that many causal sequences are more easily mastered when projected on parts of systems of pure mathematics. Such projections yield *scientific theories* (another notion used in a very wide sense!). *Language*, which makes possible most of the transfers of volition (Willens-übertragungen) that are necessary in cultural communities, lacks exactness and certainty. This holds even for the language of mathematics, which serves to evoke mathematical construction in others (though in the second part of the talk, Brouwer called the language of the mathematics of finite systems practically perfect).

In the second part, Brouwer restated, in a lucid form, the criticism of the role of logic in mathematics that he had been expounding for years. The ancients, he said, had already developed a practically perfect language for the mathematical contemplation of finite systems. In that language, there exist certain forms of transition from appropriate statements (i.e. statements indicating actual mathematical contemplations) to other appropriate statements. These forms were called the principles of identity, of contradiction, of the excluded middle, and of syllogism, and were combined under the name of logical principles. Whenever they were applied to mathematical statements purely formally (i.e. without any thought as to the

mathematical contemplations indicated by the statements) these principles held good in the sense that each new statement obtained in
this way by means of the principles was afterwards seen to evoke
actual mathematical contemplation.

What Brouwer here obviously had in mind is illustrated by the
following example. Suppose one has proved the negation of the
statement 'all elements of some finite class C have a certain property P'. If by virtue of the principle of the excluded middle one
concludes from this result that the class C contains an element
lacking the property P, then this conclusion can afterwards be tested
by inspecting the elements of C one by one; and at least one
element of C lacking the property P will actually be found.

Furthermore when applied to the language of science or to the
facts of practical life, Brouwer continued, the logical principles held
good in this same sense, the deeper reason being that mankind has
successfully mastered the majority of observable objects and
mechanisms by considering the statements about them as part of a
finite system. But from time immemorial, men have been blind to this
interpretation. Instead they have superstitiously considered words
(which are nothing but means for the transfer of human will from one
person to another) as indications of fetish-like 'concepts', and have
regarded the logical principles as laws a priori concerning the
connections between those concepts. Hence philosophers have
been confident that testable connections of concepts derived by
means of logic from undeniable axioms (i.e. connections of concepts
corresponding to undeniable facts or laws of nature) would victori-
ously pass the tests, while untestable connections so derived should
be considered with equal confidence as 'ideal truths.' Occasional
contradictions thus obtained never shook their confidence in the

logical principles but were always repaired by modifications of the axioms.

Finally mathematicians too imitated philosophers by unscrupulously applying to infinite systems the logical principles taken from the mathematical language for finite systems. In this way they derived in the mathematics of infinity 'ideal truths' that they took for more than empty words — until even in mathematics, especially in set theory, contradictions appeared, and, indeed contradictions that could not easily be eliminated by a modification of the axioms.

Hilbert's formalism tries to solve these difficulties by making the mathematical language itself the object of mathematical consideration in a language of second order and tries to give the language of first order the precision and stability of a material instrument. It completely ignores the sense (*Sinn*) of mathematical concepts and their connections and merely aims at freeing mathematical language from contradictions in a way that preserves the whole of traditional mathematics except for minor amputations which help to circumvent the known contradictions.

Intuitionism, on the other hand, Brouwer claimed, concentrates on what he called pure mathematics, outside of language. It begins by investigating to what extent the logical principles, which played such a great role in the traditional mathematics of finite systems, function for infinite systems as well. The principles of identity, contradiction and syllogism can be applied. It turns out, however, that to statements derived from the principle of the excluded middle, there corresponds, in general, no mathematical reality.

5.

That Wittgenstein was influenced by what he heard from Brouwer is clear from a letter to G. E. Moore in which Russell wrote in March, 1930, about work that Wittgenstein had recently shown him: "Then he has a lot of stuff about infinity, which is always in danger of becoming what Brouwer said." And in a report about Wittgenstein's work to the Council of Trinity in May, 1930, Russell wrote "What he says about the infinite tends, obviously against his will, to have a certain resemblance to what has been said by Brouwer."[2]

Numerous passages in the posthumously published notes demonstrate Wittgenstein's lasting preoccupation with Brouwer's ideas (if only to the extent of the contents of the lecture of 1928!); but there seems to be little in the material so far available that bears out Russell's contention of a resemblance to what was said by Brouwer.

On the contrary, in a conversation with Schlick and Waismann [3] about real numbers in December of 1929, Wittgenstein expressed the opinion that what Brouwer calls a number that is neither equal nor unequal to C (or even neither equal to C nor positive nor negative) *is not a number*, the reason being, precisely, its incomparability with rational numbers.[4] For according to Wittgenstein what is essential in the formation of a real number is that it can be compared with all rational numbers.

Wittgenstein's remark is not to be dismissed lightly. More than 10 years before it was published (though more than 25 years after it was made) Arend Heyting[5] considered the concept of real numbers that can be compared with all rational numbers, and suggests that a formalist might call such numbers *respectable*. Clearly, π is respectable in this sense. But Heyting showed by an example that the sum or difference of two respectable numbers is not necessarily respectable. One obtains a somewhat simpler example by calling β the number whose n-th decimal digit is equal to the n-th decimal digit of π except in case the latter should be the 5 in the first chromatic sequence in π. In this case, replace 5 by 4 if n is even, and 6 if n is odd. It is clear that for every rational number one can easily decide whether it is equal to, greater than, or smaller than β. But for the number $\beta - \pi$ one cannot today decide whether it is equal to, greater than, or smaller than 0, whence $\beta < \pi \to \beta - \pi$ is not respectable. One also notices that, while both π and β are comparable to all rational numbers, they are not comparable to one another. Heyting, therefore, introduces the concept only to dismiss it.

It is important in this connection that — ultimately by virtue of a deep result of Gödel's — even in the realm of real numbers that are recursively defined or computable on a Turing machine it cannot be claimed that the numbers are comparable with 0. There is no effective procedure even for deciding of such a number whether or not all its digits are zero.

Just two years after his first remark about Brouwer's numbers, Wittgenstein expressed in a conversation with Waismann (December, 1931) a completely different opinion on the subject — a change reflecting the shift of his philosophy between 1929 and 1931. He said, "There are quite diverse kinds of real numbers because there are diverse grammatical rules. Brouwer's numbers, for example, are different because the grammar of ' > ', ' = ', ' < ' for them is different."[6]

In the *Philosophische Bemerkungen*,[7] assumed to have been written around 1930, Wittgenstein mentions other decimal expansions which he, in contrast to the intuitionists, does not consider as definitions of real numbers; e. g. those that one obtains from the expansions of π or $\sqrt{2}$ by replacing all 7's by 3's. He says[8] that a real number "is an arithmetical law that yields without end (endlos) the digits of a decimal fraction." He recognizes π as a real number (without specifying which definition of π he considers as the required arithmetic law). "But one cannot produce modification of the law by operating on the decimal fraction. What one affects in this way is not the law but its accidental expression. This modification does not reach the law at all."[9]

In Part II of the recently published *Philosophische Grammatik*,[10] assumed to be written soon after the *Philosophische Bemerkungen*, Wittgenstein expounds a view that is between the opinions that he expressed to Waismann in 1929 and 1931. He now speaks about a decimal development π' identical with that of π except that where the latter contains 777 the former contains 000. In absence of a law permitting the computation of the location of 777 in π, one cannot assert either $\pi' = \pi$ or $\pi' \neq \pi$. Since the main property of numbers is their comparability, it is the main property of numbers to be comparable with other numbers . It is questionable whether or not one should call π' a number and a real number; but whatever one calls it, what is essential is that π' is a number in a different sense than is π. The possibility of a decimal expansion does not make π' a number in the sense of π.

The same is true for most of what is said about questions of this kind in Part IV of the *Remarks on the Foundations of Mathematics*, assumed to have been written in 1942 and 1943. In these notes, Wittgenstein strangely seems to have been more bothered by the assertion that a certain configuration does *not* exist in an expansion than by the assertion that it exists.

6.

A few days after his philosophical talk, Brouwer delivered a technical mathematical lecture 'The Structure of the Continuum', which Wittgenstein did not attend. In it he used a number q which he had defined in his first lecture for the purpose of disproving properties of the classical continuum of real numbers, such as linear order and connectedness. He then modified the definitions of those properties in a way that made them at least partly applicable to the corresponding intuitionistically defined sets.[11]

Two remarks lapsed into metaphysics. Brouwer emphasized that after reflection (*Besinnung*) about the primordial mathematical intuition of twoness (*Zweiheit*), the introduction of sets by his construction does not require any further reflection, nor does it represent any *petitio principii*. These facts, he said, essentially uphold the view of Kant and Schopenhauer that the continuum exists as a pure perception *a priori*, and hence independent of experience, exact and unequivocal (*unzweideutig*). Brouwer further claimed that the primordial intituion included the possibility of an interpolation between the two elements "namely, the consideration of their tie (*Bindung*) as a new element."

7.

Of course, all members of the Circle who were in Vienna at the time attended Brouwer's lectures. But his attacks on the law of the excluded middle and the consequences for mathematics of its rejection had been discussed in the Circle on several earlier occasions, his obscure remarks on primordial intellectual phenomena and primordial mathematical intuition were not taken seriously by any member of the Circle, including myself, while his voluntaristic views on communication aroused less interest in the others than they seemed to me to deserve.

Notes

[1][This visit took place in October 1927. B. McG.]

[2]*The Autobiography of Bertrand Russell 1914-1944*, pp. 292f.

[3]Friedrich Waismann, *Wittgenstein und der Wiener Kreis*. ed. B. F. McGuinness, Oxford 1967 (E.T. *Wittgenstein and the Vienna Circle*, Oxford 1979) p. 71.

[4]Actually, the number a is comparable to all rational numbers with the single exception of the number 0. But is is easy to define numbers that are incomparable to all rational numbers or to all rational numbers of an interval.

[5]Arend Heyting, *Intuitionism An Introduction*, (Amsterdam, 1956) p. 27.

[6]Waismann, op. cit. p. 188

[7]Ludwig Wittgenstein, *Philosophische Bemerkungen* Frankfurt, 1964 and *Philosophical Remarks* Oxford, 1975. [The translations from Wittgenstein in the present chapter are Menger's own. B. McG.]

[8]Ibid., p. 288

[9]Ibid.

[10]Ludwig Wittgenstein, *Philosophische Grammatik*. Frankfurt, 1969 and *Philosophical Grammar*. Oxford, 1974.

[11][See Karl Menger, 'Intuitionism', in *Selected Papers*, for a fuller history of intuitionism and its relation to more traditional views of mathematics. L. G.]

XI. DISCUSSIONS IN THE CIRCLE 1927-30

Most of what Karl Menger had to say about the discussions within the Vienna Circle was published in his 'Introduction' in Hans Hahn, Logic and Mathematics, ed. McGuinness, Dordrecht, 1980, pp. XI-XVII, and 'Memories of Moritz Schlick' in Science and Rationality, ed. E. Gadol, Vienna & New York, 1983. There are echoes of what he says there in the sketches given earlier in this volume. There is also discussion of the subject in his 'Memories of Kurt Gödel,' which is included in this volume. However, it appears that he may have been planning a complete chapter on this subject. We here print the scant material we have found concerning the period before Menger's visit to Poland.

Menger had some differences of opinion with members of the Circle. The first arose from the contention that mathematics was a system of tautologies. Menger continues as follows:

So I finally arrived at the conclusion that the members of the Circle must be thinking of the reduction of mathematics to logic claimed in the *Principia Mathematica* in 1910. That mathematics proceeds according to logic has of course been known for 2000 years. That logic reduces analysis to arithmetic is one of the great results of the nineteenth century. What Russell added in his claimed reduction was a definition of the arithmetical terms, and a proof of the arithmetical axioms, by logic. This last step in the reduction was disputed by Poincaré and by the intuitionists and has ceased to be unequivocal because of the discovery of n-valued logics, all of which came up in the Circle later. But be that as it may, I failed to see what the ill-defined use of the term 'tautology' in the provocative reference to mathematics as a system of tautologies substantially added to Russell's claim that mathematics can be reduced to logic.

Another of my questions concerned language. I objected to the recurring references in the Circle to *the* language and repeatedly asked Carnap, Schlick and other members what justified the implied belief in the uniqueness of language. But on this point, too, I failed to receive a satisfactory answer. Schlick did not seem to take the question seriously.

In the course of the following years, however, Carnap not only gave up that belief but emphasized the importance of the existence of a multiplicity of languages between which one may choose, while Schlick and Waismann continued speaking about *the* language.

In 1937, when his Viennese period was still fresh in his mind, Carnap wrote in his book *Logical Syntax of Language* ". . . the earlier position of the Vienna Circle, which was in essentials that of Wittgenstein. On that view it was question of *'the* language' in an absolute sense; it was thought possible to reject both concepts and sentences if they did not fit into *the* language."[1] Carnap wrote even more explicitly that same year in *Testability and Meaning*: "But I was wrong in thinking that the language I dealt with was *the* language, i.e. the only legitimate language, — as Wittgenstein, Schlick, and Lewis likewise seem to think concerning the language-forms accepted by them."[2] In contrast to these passages, in the Intellectual Autobiography, written twenty five years later, Carnap strongly emphasizes that the idea of freedom to choose one of a variety of languages had already been one of his leading thoughts in his pre-Vienna period. The development of the idea about the multiplicity of language is one of those facts that Carnap seems to have completely misremembered in his later years.

Notes

[1]Rudolf Carnap, *The Logical Syntax of Language*, Patterson, New Jersey, 1959, p. 322. [1937 saw the first publication of the English translation. The German original (*Logische Syntax der Sprache*) was published by Springer, Vienna, in 1934, but this hardly detracts from Menger's point. B. McG.]

[2]Rudolf Carnap, 'Testability and Meaning', *Philosophy of Science* v. 4, no. 1, January, 1937, p 20.

XII. POLAND AND THE VIENNA CIRCLE

1.

After I had sent offprints of my first paper on dimension theory to the Warsaw topologists in 1923, we regularly exchanged offprints of our publications. About 1920, the Warsaw mathematicians had started a remarkable journal, the *Fundamenta Mathematicae*, unique in its devotion to abstract set theory, the theory of point sets, and the foundations of mathematics. Soon I was in correspondence with several contributors, especially Bronislaw Knaster and Kazimierz Kuratowski. In the autumn of 1929, the mathematicians of the University of Warsaw invited me there to deliver a lecture.

As I observed during this and subsequent visits, Warsaw between the two world wars had a marvellous scientific atmosphere. The interest of the mathematicians in their own as well as their colleagues' and students' work was of an intensity that I have rarely observed in other mathematical centers. I discovered the same spirit in the Warsaw School of Logic. But up to that time the Polish logicians had been somewhat isolated.

The country to which Polish scholars traditionally turned in order to attract the attention of the Western world was France, with which Poland was also in a close political alliance. Polish mathematicians published many of their results in the *Comptes Rendus* of the Paris

Academy of Science, but in the field of logic the situation was somewhat different. The French logicians Louis Couturat and Jean Nicod were no longer active; and the brilliant young Jacques Herbrand could not help the Polish logicians much in France. In fact, before his untimely death a few years later, he published some of his own most important papers in Polish journals.

French mathematicians were not greatly interested in mathematical logic: Poincaré had derided Russell's derivation of arithmetic from logic; and the great set theorists and analysts of the Paris school were in disagreement about foundational questions. In fact, when a young Polish logician submitted an interesting logical paper dealing with the axiom of choice to one of the most outstanding mathematicians of France the paper was rejected on the grounds that Zermelo's axiom was invalid and hence the result of the paper false; and when, undaunted, its author submitted the paper to another eminent French mathematician it was again rejected, this time on the grounds that the axiom of choice was evident and hence the result trivial.

In England, in the 1920's, some but not many of the Polish results were known. All that Whitehead and Russell quote in the second edition of the *Principia Mathematica* (1927) is Leon Chwistek's work on the antinomies and three papers by Alfred Tajtelbaum-Tarski, published in *Fundamenta Mathematicae*.

In Germany, in the 1920's, Abraham Fraenkel was familiar with Polish set theory but was less versed in Polish logic; the logicians in Göttingen were not yet fully familiar with the results obtained in Warsaw; nor had the relations of the Polish logicians with Heinrich Scholz and his group in Münster yet developed. The majority of Germans were intensely hostile to the restored Polish nation because of the loss, in the peace of Versailles, of the territories inhabited by Poles, especially the so-called Polish Corridor, which joined Warsaw to the sea while separating Berlin from Königsberg, the city of Kant. Even many German intellectuals had an idiosyncratic aversion to Poles, which the latter, mindful of one hundred and fifty years of oppression by Prussia, reciprocated.

Vienna and the Viennese, on the other hand, before the rise of Hitler were well-liked by Polish-speaking people, gentiles and Jews alike. Of the three neighbor nations that had divided up the Polish kingdom in the eighteenth century (Russia, Prussia and Austria), only Austria had treated the Poles with consideration; and even when German nationalism developed in Vienna there was relatively little friction with the Poles since the province of Galicia which they inhabited had (in contrast to the Czech territory) no extensive common border with the German parts of the empire. Almost every Imperial Austrian government had included a Polish member, on one occasion as prime minister. In Galicia, the elementary and secondary schools were Polish. The University of Cracow, the alma mater of Copernicus, (founded in 1364, a year before the University of Vienna and, next to that of Prague, the oldest university in Central Europe) flourished; and under Austrian rule, a University of Lwow (Lemberg) and a Cracow Academy of Sciences were founded. Besides cultivating Polish traditions, these institutions developed a many-faceted Polish culture distinguished by remarkable scholars, especially mathematicians, physicists, and philosophers. Rationalism and logic pervaded Polish intellectual life.

So even those who left Galicia after World War I and went to Warsaw, the capital of the liberated nation, kept rather pleasant memories of the old Austria. As for example, Waclaw Sierpinski one of the founders of the Polish school of mathematics, sent me from

Warsaw a photograph of the decree appointing him to his first professorship, signed by the Austrian minister of public instruction.

2.

The first Polish logician I met, near Cracow, on my way to Warsaw was Chwistek and we soon found points of agreement about obscurities in Husserl's phenomenology and Weyl's utterances in support of it.

Stanislaw Lesniewski, who had been described to me as something of a Polish Wittgenstein, had just published a very verbose account of his ideas in *Fundamenta Mathematicae* (v. 14). Unfortunately I barely met him. For when he saw me he exclaimed in honest surprise "Are youngsters made professors in Vienna?" I felt greatly amused and flattered since I seemed to have successfully camouflaged my receding hair line. But my mathematician hosts were very angry, apologized in front of Lesniewski for his remark, which despite my protest, they considered as an insult to the University of Vienna and to me, and dragged me away. Unable to win them over to my way of looking at this trivial matter, I never saw Lesniewski again.

Lukasiewicz, extremely formal almost to the point of stiffness, was personally the exact opposite to Lesniewski. In 1929 he was in Lwow; and I first met him during a subsequent visit to Poland. But I learned about his important logical achievements (some of which will be discussed in this chapter) as soon as I came to Warsaw from Alfred Lindenbaum, Tarski, and other young logicians.

It was not only the formal logical work of the Polish logicians, however, that impressed me. I also noted that they were interested in philosophical problems similar to some discussed in the Vienna

Circle, but they attacked them in connection with, and partly on the basis of, their exact logical studies. They always confined themselves to *concrete* questions and completely eschewed those vague generalities which seemed to me to becloud some of the Vienna discussions in the late 1920's.

So I decided to familiarize the Vienna Circle as well as the members of my Mathematical Colloquium with the logico-philosophical work of the Warsaw school and invited Tarski to deliver three lectures before the Colloquium, to two of which I planned to invite also the entire Circle. Tarski eagerly accepted.

3.

Tarski delivered his three lectures before the Colloquium[1] on February 19, 20, and 21, 1930. The first was devoted to set theory and for the other two Tarski had asked me to make a selection from several logical topics. I chose two which not only illustrated the work done in Warsaw but seemed to me also to fill the needs of the Circle. I announced the two lectures to its members in the meeting prior to Tarski's visit adding a cordial invitation and a description of my impressions in Warsaw. Unfortunately, however, on February 20 only Hahn and Carnap came to join the Mathematical Colloquium.

The lecture dealt with some fundamental concepts of the methodology of the exact sciences using *meaningful proposition* and *consequence* as undefined terms. If T is any set of propositions, let $Cn(T)$ be the set of all consequences, and S the set of all propositions. The following four meta-propositions served as axioms: 1) the set S is denumerable; 2) for each set T of propositions, $Cn(T)$ is a subset of S containing T as a subset; 3) $Cn(Cn(T)) = Cn(T)$; 4) $Cn(T)$ is the union of the sets $Cn(X)$ for all subsets X of T. A set D of

propositions is called a *deductive system* if Cn(D) = D. A set G of
propositions is *axiomatizable* if there exists a finite set of propositions
F such that G = Cn(F). Lindenbaum discovered nonaxiomatizable
sets of propositions.

In selecting the topic of the lecture, I had hoped that it would
convince the Circle of the possibility of obtaining valuable, if not
necessarily earth-shaking, methodological results from unproven propositions in terms of undefined words, developed in what Hilbert
called a metalanguage — a possibility that Hahn, unfortunately without success, had been emphasizing in the Circle all along. Carnap
soon recognized the significance of the topic for philosophy and
expressed his appreciation. But since, besides Carnap, only Hahn
had joined the mathematicians in the Colloquium that evening there
was no one who had to be convinced and there was very little
discussion. I was greatly disappointed and displeased.

So, before Tarski's last lecture, I invited by individual phone calls
all members of the Circle to attend with great — perhaps too great —
urgency. (Unfortunately, I have always lacked the patience to be
diplomatic in order to convince people to accept what, rightly or
wrongly, I feel is offered for their own good.) Be that as it may, on
that last evening the Colloquium was joined by the full Circle.

4.

Tarski discussed truth tables and similar devices used to define the
logical particles explicitly, and systems of axioms used to introduce
them implicitly. As an example, he mentioned the following system of
three independent axioms due to Lukasiewicz (but different from

those listed in VII,3):

(α) (p implies p) implies ((q implies r) implies (p implies r));

(β) p implies (not p implies q);

(γ) (not p implies p) implies p;

The operation (I) of Modens Ponens and (II) of Substitution, which are used in connection with axioms (1), (2), (3) are also employed to derive the entire calculus of propositions from (α), (β),(γ).

Incidentally, the tautology (γ) corresponds to a mode of reasoning used In Euclid's *Elements* and later by Girolamo Saccheri, the precursor of noneuclidean geometers, namely,

(III) If the proposition (not-P implies P) is admitted, then so is P.

Indeed, in order to prove a certain geometric proposition, Euclid and Saccheri proved that not-P implies P. Conversely, Modus Ponens corresponds to a tautology, namely to the tautology

(p and (p implies q)) implies q.

But if this tautology were added to (1), (2), (3) or to (α), (β), (γ) as an axiom, then the single procedure (II) would by no means yield the entire calculus of propositions.

Tarski futher discussed Lukasiewicz' 3-valued logic dividing propositions into three classes, which I shall denote by T, ?, and F. The truth tables for the particles *implies* and *not* follows. The four lines (marked by arrows) that include only T's and F's are identical with the corresponding lines in the corresponding tables for the 2-valued logic (see VII,3).

p	q	p implies q
T	T	T
T	?	?
T	F	F
?	T	T
?	?	?
?	F	?
F	T	T
F	?	T
F	F	T

p	not p
T	F
?	?
F	T

I asked Tarski to include in this talk a remark about Lukasiewicz' *frontal*, *parenthesis-free notation*, which I had become familiar with in Warsaw. Instead of symbolizing the phrase 'p implies q', in the traditional way by a sandwiched sign, e. g. often an arrow as in p → q, Lukasiewicz wrote Cpq. This frontal symbol, *C*, makes parentheses or brackets unnecessary. The expression, p → q → r, is ambiguous and parentheses are indispensable in order to distinguish

p → (q → r)   and   (p→ q) → r.

Instead Lukasiewicz wrote, unambiguously and without any need for parentheses,

CpCqr   and   CCpqr.

I thought that Lukasiewicz' method would interest the Circle since Wittgenstein had written in the *Tractatus* 5.461. "The apparently unimportant fact that the logical pseudo-relations like V and ⊃ (*or* and *implies*) need brackets — unlike real relations — is of great importance. The use of brackets with these apparently primitive signs shows that these are not real primitive signs; and nobody of course would believe that the brackets have meaning by themselves.". In the discussion following Tarski's lecture, however, Waismann rose with a rather inane defense of Wittgenstein. But even Schlick nodded assent when Hahn remarked in a somewhat irritated tone "But Mr. Waismann, why not admit that on this point Wittgenstein was evidently mistaken?" Then Carnap expressed his appreciation of both talks on logic from philosophical point of view. His words, drew mixed comments from members of the Circle. But the lecture was received with great interest by Hahn as well as Gödel and the other mathematicians of the Colloquium. Neurath seemed favorably disposed, Schlick a bit cool — partly, I think, because of the over-urgency of my invitation — and Waismann was totally unimpressed.

5.

After Tarski's departure, I continued to think about Lukasiewicz' notation. First of all, it is obvious that his method would be applicable to arithmetic and algebra, where the sandwiched symbols + and ·, for addition and multiplication might be replaced by frontal *S* and *P* for *sum* and *product* as in *S*23 and *P*23 for 2 + 3 and 2 · 3. The distributive law of algebra could then be written without obnoxious parentheses,

PaSbc = *SPabPac* instead of a·(b + c) = (a ·b) + (a · c).

(Of course, an expression such as *S*2347 would be incomprehensible unless separators, such as commas or spaces, were used in order to distinguish

$$S2,347 \quad S23,47 \quad S234,7.$$

Forty years later, in the age of computing machines, all this is of course common knowledge.)

It occurred to me that the introduction of frontal notation would in fact have been more natural for the Renaissance mathematicians in arithmetic than it was for Lukasiewicz in logic. If the founders of traditional arithmetic notation had only followed the words of everyday language, 'the sum of 2 and 3', they would have written $S23$. Since on the other hand, everyday language lacks a phrase

implication p q   for   p implies q,

it required greater ingenuity on the part of Lukasiewicz to invent the frontal notation for logic.

Not that the founders of traditional notation were indifferent to the obnoxiousness of parentheses! After introducing sandwiched symbols, which make parentheses indispensable, they tried to get rid of parentheses by various tricks, such as the convention, *first multiply, then add*.[2] After adopting this rule one can write 2 + 3·4 instead of 2 + (3·4), while the parentheses in (2 + 3)·4 remain of course indispensable. In frontal notation one could write $S2P34$ and $PS234$ without parentheses in either case.

On closer inspection, the traditional notation reveals other undesirable features. It is unsystematic in character, in that the arithmetical symbols are of four types to which a fifth type is added in logic and advanced mathematics:

    a) *sandwiched*, as in 2 + 3;

    b) *frontal*, as in $\sqrt{2}$ and log 2;

    c) *rear*, as the raised 3 for cube in 2^3;

    d) *pincers*, as the two bars for absolute value in $|-2|$.

e) *interlaced* as in *neither . . . nor*, and *∫ . . . d. . .*

Not only sandwiched symbols but also combinations of frontal and rear symbols necessitate the use of parentheses, all of which could be avoided if frontal symbols only, such as *cu* for cube, were used. One could then write *log cu2* instead of $(log2^3)$ and *cu log2* for $(log2)^3$.

The symbol $log2^3$ is of the same ambiguity as the expression 'the father of Henry's mother.' In fact all types of arithmetical symbols have analogues in descriptions of family relations:

a) *sandwiched*, as in Henry's daughter by Anne;

b) *frontal*, as in the daughter of Henry:

c) *rear*, as in Henry's daughter;

d) *pincers*, as in the oldest of Henry's children.

e) *interlaced*, as in the daughter of Henry by Anne.

In common language, there seems to exist a healthy trend toward frontal expressions.

I later found deficiencies in the symbolism of pure and applied analysis — deficiencies beginning in calculus — that were much more serious than those in arithmetic and algebra, and I worked to correct them. But it was the dispensability of parentheses which first made me aware of the shortcomings of traditional mathematical notation.

6.

Strings of meaningful words may be meaningless because of the mere arrangement of the words, in particular, because of the distribution of logical particles. Husserl and Carnap gave the examples:

Green is or   and   *And or of which*.

Compound propositions are meaningful if and only if they are built up from component propositions (indicated by letters p, q, ...) in a certain way; which leads to what is today called a *well-formed expression*. The procedure is as follows:

(1) Single letters are examples of well-formed expressions. So are particles such as 'and', 'or', 'implies', sandwiched between two letters, and 'not' followed by one letter.

(2) If a letter in a well-formed expression is replaced, by a well-formed compound, the result is well-formed. For example, since 'p implies p' is well-formed, so is 'p implies (q implies r).' In contrast, the substitution operation (II) in the Calculus of Propositions requires that a letter be replaced by the same expression in *each* of its occurrences.)

For a very long string of symbols, however, it is rather difficult to ascertain whether or not it can be generated in this way. A few months after Tarski's visit it occurred to me that well-formed expressions in frontal notation can be characterized by simple arithmetic criteria.[3]

Let **S** be a string consisting of (lower case) letters in roman type, p,q, . . . indicating propositions, and two kinds of letters in *italic type*: *capitals* such as frontal C in Cpq for p 'implies', and other capitals as frontal indications for 'and', 'or', and other particles connecting two propositions; and *lower case* letters such as a frontal *n* in *n*p for 'not' p. I proved that **S** is well-formed if and only if

I. *The number of Roman type letters in* **S** *exceeds the number of capitals exactly by 1.*

II. *Each initial segment of* **S** *(except the entire* **S** *) contains at least as many italic capitals as Roman type letters.*

III. *The last element of the string is not a lower case italic letter.* (It readily follows that the last letter in **S** is in Roman type).

One can easily verify these conditions for the axioms of the calculus of propositions mentioned in VIII,3), when written in frontal notation.

(a) *CCpqCCqrCpr*,   (β) *CpCnpq*,   (γ)*CCnppp*.

The string (a) includes 6 lower case letters and 5 *C*'s while each initial string contains either just as many lower case letters as *C*'s, e.g. *CCpqCCqr*, or more *C*'s than lower case letters, e.g. *CCp* or *CCpqC*.

For two strings, the first being (a), we give under each element the numbers of capitals and of Roman type (r.t.) letters up to that element:

	C	C	p	q	C	C	q	r	C	p	r
Cap.	1	2	2	2	3	4	4	4	5	5	5
r.t.	0	0	1	2	2	2	3	4	4	5	6

	C	C	C	p	q	r	C	s	t	C	C	u	v
	1	2	3	3	3	3	4	4	4	5	6	6	6
	0	0	0	1	2	3	3	4	5	5	5	6	7

In the first example, the figure in the Cap. row is nowhere less than that in the r.t. row. In the second example, up to t there are 4 Cap.'s and 5 r.t.'s. Hence the second string is not well-formed.

What interested me in this matter was the fact that *by merely carrying out a simple procedure one can decide whether or not an expression satisfies a necessary condition for meaningfulness.* I mentioned that fact to Carnap, who had just then begun to be interested in syntax.

If the letters in Roman type indicating propositions are replaced by numerals or Roman type letters indicating numbers, if the italic capitals are *S, P* for sum and product or other frontal symbols for binary operations and if the lower case italics indicate unitary operations such as *m*, for minus, or *rec* for reciprocal or *log*, then arithmetical or algebraic expressions result, which are well-formed if and only if they satisfy Conditions I, II, III. That well-formedness is necessary but not sufficient for meaningfulness is demonstated by the expression *Q*00, where *Q*ab indicates the quotient a/b.

7.

During Tarski's stay in Vienna, I arranged an opportunity for Gödel to give the visitor an account of his thesis, whose result greatly impressed Tarski. Moreover, Carnap and Tarski had a long discussion, in the course of which Tarski invited Carnap to deliver a lecture in Warsaw. Carnap went to Poland in the spring of 1930. Carnap's relations with the Polish logicians were to have a considerable influence on his development; and Tarski's visit in Vienna was certainly one of the first important steps out of isolation for Polish logic and philosophy. At the station before his departure, Tarski thanked me enthusiastically for the days he had been my guest, although I emphasized that all I had tried to do was to further the common interests of two scientific-philosophical groups. Yet he embraced me and said he would never forget what I had done for him.

Notes

[1]*Ergebnisse Mathemat. Kolloquiums* 2 (1930) p. 12. The remark in Carnap's autobiography that Tarski came to Vienna at the invitation of the Mathematics Department is incorrect.

[2]In another culture, mathematicians might stipulate *first add, then multiply*. Then 2 + 3 • 4 would stand for (2 + 3) • 4 and one would have to write 2 + (3 • 4). The arithmetic and algebraic formulas would look quite different from ours while expressing the same contents.

[3]Published in the *Ergebnisse*, Heft 3. When I saw Lukasiewicz in Warsaw in the early 1930's he told me that these conditions had also been discovered by Stanislaw Jaskowski in Poland.

XIII. THE UNITED STATES 1930-31

1.

I lectured on dimension theory and metric geometry at Harvard University from September 1930 until February 1931. While I was there I also tried to learn about philosophical studies parallel to those carried on in Vienna. However, while I had interesting mathematical conversations, especially with George David Birkhoff, the outstanding American mathematician, and Marston Morse, few people seemed to know much about the Vienna Circle. This changed with the arrival of Feigl, who began to propagate Viennese ideas and especially Carnap's wherever he went.

Birkhoff was intensely interested in aesthetics at that time. He had just returned from a world tour and brought home interesting phonograph records of various types of Oriental music. While the experience certainly had not made him an aesthetic relativist, there were some positivistic elements in his aesthetic studies. Steering clear of what he called 'philosophical definitions' of art (such as Croce's 'the expression of an impression') Birkhoff *tested* his theory of beauty. He let groups of students rank polygonal figures according to beauty and compared the results with the ranking according to his formula. The latter was based on the idea that, roughly speaking, beauty increases with order and decreases with complexity. In fact, he declared the ratio of order to complexity as the measure of beauty without, in my opinion, sufficiently stressing the arbitrary character of this stipulation. Of course, he was aware of the need for numerical evalutions of order and complexity as a prerequisite for a discussion of their quotient; but in assigning such values he could not help introducing numerous further elements of arbitrariness. Certainly his result was far from reflecting my own aesthetical feelings. It suffices to say that highest in Birkhoff's order of beauty of polygons stood a square and a rectangle with sides parallel to the edges of the paper (the longer sides of the rectangle being horizontal for the sake of 'stability'), and an equilateral triangle with a horizontal base.

2.

Of the Harvard philosophers, the one I saw most frequently was Henry Maurice Sheffer, a very personable and likeable man. On hikes in the beautiful surroundings of Cambridge we had long, if inconclusive, talks about philosophy. Does logic describe correct thinking and what is *correct* thinking anyway? Or is logic simply the grammar of the particles *not, and, all,* etc.? Is it a coincidence that practically all languages agree[1] in the use of the logical particles?

Sheffer had discovered in 1913 that all particles of the calculus of propositions (see VIII,3) can be expressed in terms of a single one:

in terms of p *incompatible with* q (i.e., *not both* p *and* q),

as well as

in terms of *neither* p *nor* q (i.e., *not* p *and not* q).

For example,

(*not* p) is equivalent to

(p *incompatible with* p) as well as to (*neither* p *nor* p).

(a)

(p *and* q) is equivalent to

(p *incompatible with* q) *incompatible with*

(p *incompatible with* q).

(β)

(*If* p *then* q) is equivalent to

(p *incompatible with* q) *incompatible with* p.

In frontal notation, one might write Apq, Bpq, Cpq for p *incompatible with* q, p *and* q, p *implies* q, respectively. Then (a) and (β) read

$$Bpq = AApqApq \text{ and } Cpq = AApqp.$$

In the *Tractatus*, Wittgenstein made extensive use of Sheffer's discovery without mentioning its author. In one of our conversations, Sheffer brought up this fact adding (not quite convincingly) that he did not resent it. (In the Preface to the *Tractatus* Wittgenstein wrote ". . . what I have written makes no claim to novelty in points of detail; and therefore I give no sources . . ." "Russell in his Introduction to the *Tractatus* gave me full credit," Sheffer said. Then he spoke about the importance for him of Russell's later support in view of the speed with which gossip spreads in universities. I listened with surprise since I was under the misapprehension that he was still talking about the *Tractatus*. Only decades later, while reading Russell's *Autobiography*, did I realize that Sheffer was probably alluding to the fact, mentioned in that book, that a few years before my visit he had experienced serious social difficulties at Harvard. If they still existed in 1930, he was certainly clever enough to ignore them and had enough sense of humor to laugh at those who were prejudiced against him. But in the early 1920's, according to that *Autobiography*,

Sheffer without Russell's strong support might have lost his job. In that situation, professorial gossip might well have kept him from obtaining another position commensurate with his attainments.

3.

Paul Weiss was preparing an edition of Charles Sanders Peirce's *Collected Works* with Charles Hartshorne. Weiss told me a great deal about that profound logician, who in the 1920's was not well known in Vienna, where pragmatism was regarded as having been originated by William James, although James himself gave the credit to Peirce. Peirce was the first to stress "the consideration of the practical bearing that we conceive the objects of our conceptions to have" and wrote in a frequently quoted passage that "our conception of those effects is the whole of our concepts of the objects."

As far as I could see, Peirce was not articulately antimetaphysical. In fact, he said that metaphysics has to study the most general features of reality and of real objects. But besides formulating the theses of pragmatism he took, in a less frequently quoted passage, a remarkable step toward a critique of language. "Whatever statement you make (to an experimenter)," he wrote, "he will either understand it as meaning that if a given prescription for an experiment . . . is carried out. . ., an experiment of a given description will result, or else he will see no sense in what you say."

Many manuscripts of Peirce's anticipating important work by later logicians, especially Sheffer's discovery and parts of the theory of relations remained unpublished during Peirce's lifetime. But even the papers on logic that he published attracted little attention.

"That the mere external exigencies of Peirce's life and the indifference of publishers prevented any full-length presentation of his philosophy," the editors wrote in their Introduction to Peirce's Works, "is a tragedy." I thought that Peirce had been professor at Harvard. "No," Weiss told me, "His father and brother were. He himself was on the faculty of Johns Hopkins and only for a short time. People did not like his morals." I thought of the loss that American universities inflicted upon themselves at the turn of the century by rejecting the services of one of the greatest American thinkers.

4.

I often saw Norbert Wiener, who was associated with the Massachusetts Institute of Technology, and with him, too, took long Sunday hikes in the beautiful surroundings of Cambridge. A biographer has mentioned Wiener's feeling of insecurity. This feeling probably explains his constant need for the recognition of his achievements including his linguistic proficiency. While in most cases, I agreed with Wiener's self-assessment I found it somewhat boring to profess my agreement several times in each conversation. But then I decided to look at the amusing aspect of his foible and counted how often in each of our otherwise interesting conversations, I had to assure him of the excellence of his German, his French, etc.

As readers of Wiener's autobiography know, not much love was lost between him and Birkhoff. But when he started this topic I told him that I had always found that excellent mathematician very pleasant and that he had been instrumental in arranging my visit to the United States. Wiener spoke more kindly of Russell with whom he had studied as a very young man before and during the first world war in England. As a result, he published several logical and philosophical papers during that time. Later, he wrote the article 'Aesthetics' in the *Encyclopedia Americana*; and even a strict positivist can subscribe to most of what he said. Owing to the *de facto* uniformity of our nature, certain types of things will, all in all, attract us in the long run while others will repel us. This fact is, Wiener said, what one expresses by saying that we have a *taste*, which changes slowly, if at all. But he is, in my opinion, too vague in saying that according as we apply our taste to an object 'in its environment and in its causal chain' or 'in complete isolation', we arrive at moral or aesthetic evaluations. After this overgeneralization, however, Wiener returned to sound descriptions. "Taste with a capital T is formed from the aggregation of individual tastes." But there are always certain individuals whose have been more developed than those of the others and who indicate the direction that the general human taste will assume with the increase of education. "The norm of aesthetics . . . is to be found in the direction in which evolution and education are leading the judgments of mankind. It is because of the *de facto* permanency of the trends of education that art is a social matter . . and not because of any *a priori* character of the beautiful and of art."

In 1930, Wiener was in his post-logical, pre-cybernetic period, entirely preoccupied with mathematical analysis. Whatever mathematical problem was being discussed and even in sundry non-mathematical discussions, Wiener would bring up a more or less relevant Fourier integral.

5.

I had many conversations with Edward Huntington, a very kindly person. He had axiomatized several elementary mathematical theories, and his work was not only systematic and reliable but in 1902, it brought out an important new idea that is also of considerable philosophical interest and was often mentioned in the Vienna Circle, especially by Carnap. It was the idea (often attributed to O. Veblen) of *a postulational theory that holds essentially for only one set of entities*; more precisely, the idea of a system of axioms such that any two sets of entities satisfying the axioms are *isomorphic*. (Isomorphism of two sets, a concept of paramount importance not only throughout mathematics but also in the philosophy of science, means the existence of a one-to-one correspondence between their elements such that all relations dealt with in the axioms that hold for elements of one set also hold for the corresponding elements of the other.) In 1902, Huntington called such systems of axioms *sufficient*, the idea being that the system is sufficient to characterize essentially *one* set of interrelated entities, whereas a part of those axioms would in general be satisfied by *many* sets and therefore be *in*sufficient to characterize any one. Huntington's sufficient systems of axioms were later called *categorical* by O. Veblen and *monomorphic* by Carnap.

The system consisting of the axioms I, II, III is *poly*morphic since it admits many mutually nonisomorphic realizations; e.g. one consisting of 4 points and 6 lines, another one consisting of 9 points and 12 lines, etc. But if one adjoins the axiom

IV. *There are at most 4 points*,

then the system consisting of axioms I-IV is monomorphic. In fact, it is easy to show that any set of elements called points and lines, satisfying I-IV is isomorphic to the vertices of a tetrahedron.[2]

One of the topics axiomatized by Huntington was the arrangement of the points on a straight line. The (undefined) points are assumed to be either in a binary relation: P is *less than*, or *to the left of* or *below* q, or (following Pasch) in a ternary relation: Q is *between P and R*, both these relations being undefined and subject merely to certain assumptions about properties of theirs, e.g. if Q is between P and R, then 1) Q is between R and P; 2) R is not between P and Q; 3) if also R is between P and S, then Q is between P and S, and R is between Q and S; 4) if Q is between P and R, and R is between Q and S, then Q is between P and S.

This theory of linear betweenness was so firmly established that even philosophers were interested in my formally and substantially different theory, which is likewise applicable to the world, and Huntington asked me about it. Its framework is a metric space with undefined points every two of which have numerical distance from one another. The theory is *formally* different from that of linear order in that the between relation here is *defined*, namely by saying that Q is between P and R if the distance PR is equal to the sum of PQ and QR. It is *substantially* different in that the relation so defined shares some properties with linear betweenness, e.g. 1), 2), 3), but may lack others, e.g. 4).

Consider, for example, in the railroad plane whose points are stations, while the distance between two points is the time duration of the shortest train ride between them. Q is between P and R if Q is on the shortest route connecting P and Q (or on one of the shortest routes if there are several routes that are equally long but shorter than all the others). In 1930, there existed two equally shortest train connections of New York and Chicago, one via Cleveland and the other via Pittsburgh. These cities were thus between New York and Chicago. Youngstown was between Cleveland and Pittsburgh (on a North-South line connecting these two cities) but not between New York and Chicago since there was no shortest train connection of the metropolises via Youngstown. That some further properties of linear betweenness may break down for metric betweenness can be seen in the taxi cab plane. In Fig. 2a, all points in the streets within the rectangle having O and P as opposite corners are between O and P.

6.

Other logicians and philosophers at Harvard included Alfred North Whitehead and Willard V. O. Quine. The former, with whom I exchanged only a few words at one of his Sunday teas, was 70 but appeared to me even older and seemed to have lost his interest in logic. Quine, on the other hand, while obviously destined to become an outstanding logician, was still working for his PhD. Unfortunately I did not know that C. I. Lewis had joined the faculty that fall and therefore made no attempt to meet him.

7.

The greatest impression that I received from a philosopher of science at Harvard or, for that matter, in the United States came from Percy Bridgman, a quiet and amiable man with a wiry figure. In conversations and debates he always watched the speaker closely and critically; then after a moment of deliberation, he replied briefly and pointedly. A glint would appear in his eyes when the conversation took a turn that he seemed to have anticipated or desired. He had written that between men schooled in operational thinking much conversation will become unnecessary and that the ultimate effect of this schooling would be to release one's energies for more stimulating and interesting interchanges of ideas. Conversations with Bridgman were stimulating and interesting indeed.

Bridgman appeared to me as a modern reincarnation of Mach. Both men were experimentalists with an encyclopedic knowledge of the physics of their time, who devoted a good deal of their energy to reflections about the basic concepts and the structure of physics, realizing the importance of such studies for physics and in fact for science in general.

At Harvard, I became acquainted with Bridgman's *The Logic of Modern Physics*, published in 1927[3] and immensely enjoyed reading parts of the book. At the exact time when Schlick defined *the meaning of a statement* as *the method of its verification* Bridgman wrote "the proper definition of a concept is in terms of actual operations, in general, we mean by a concept nothing more than a set of operations; *the concept is synonymous with the corresponding set of operations*," the concept of length, for instance, with the set of operations by which length is measured and determined. His and Schlick's points of view obviously supplement and compliment one another. But while the philosophers in the Vienna Circle aimed primarily at general statements about procedures, Bridgman stressed particulars. In fact, he emphasized the differences between procedures yielding different concepts or even the same concept in different ranges, for "as phenomena change in range their character may change." His encyclopedic knowledge enabled him, as far as physics is concerned, to penetrate into almost all important particulars.

As Bridgman repeatedly said, his point of view was largely inspired by the basic idea of the special theory of relativity. There is also a similarity between the sentences from his book quoted above and certain passages in Peirce's writings. Some of what Bridgman said about mathematics in the book was weak; but he was eager to improve his views. If I remember correctly, Bridgman was one of those who first learned about the Vienna Circle from Feigl. He was quite interested and asked me about several details concerning the

Circle as well as intuitionism. In the matter of relativity of rotations, Bridgman emphatically adopted and further developed Mach's view.

8.

In the fall of 1930, Josef Schumpeter was teaching in the Harvard Economics Department. He was a man of great personal charm, a bit affected but very entertaining. He is not usually mentioned in connection with philosophy. Yet Schumpeter was, I believe, the first to introduce, as early as 1908, Machian ideas into economic theory in his book *Essence and Content of Theoretical Economics*.[4] In particular, he asked for descriptions rather than explanations and tried to replace cause-and-effect relations by functional connections.

In the second generation of Austrian marginalists, Schumpeter was the only one without prejudices against mathematics. In fact, he had some knowledge of mathematical analysis, which he was very eager to expand. For he had a strong wish to understand and to follow the work of the young mathematical economists whose numbers rapidly increased in the late 1920's.

In 1930, Schumpeter's mathematical knowledge was still rather primitive. But he diligently studied differential equations — I believe from one of the French treatises — and told me about his progress whenever we had lunch in the Harvard Club. One day, when he had just reached the chapter on existence proofs in his treatise, he spoke about their solution with great indignation. "In case of those differential equations," he said, "one has even to worry whether the bitch exists (ob das Luder existiert)." I tried to explain that such worries must not be confined to *differential* equations. Any equation results from a certain statement about a connection between some entities and numbers, and consists in the task of 'finding' some of those entities when others are 'given'; in other words, it is the problem of expressing some of them in terms of others; and indeed of numbers mentioned in the statement. But, in general, it is uncertain whether the basic statement is valid for *all* combinations of 'given' entities and numbers or in fact for *any* such combination; and if this is not the case, then the 'bitch' does not exist.

But Schumpeter did not seem to take these remarks very seriously, not even when I mentioned, as an example, the socalled problem of imputation (*Zurechnungsproblem*) of the marginalists. In contrast to the eighteenth century economists, who taught that the value of products stems from the value of the factors of production, the marginalists claimed that, the value of the factors is derived from the expected value of the products. The question arose as to how one imputes the prices of the products to the factors. Walras had set up a system of equations for the prices and quantities of the factors used in the production or, rather, he had set up statements connecting the prices and quantities of the factors and of the products and one could try to express those of the factors in terms of those of the products, "No differential equations are involved," I said to Schumpeter, "but are you sure that a solution exists?" Schumpeter brushed this off with the remark that Walras' system included just as many equations as unknowns. "So does the problem of finding two numbers x and y satisfying the two equations $x + 2y = 3$ and $2x + 4y = 5$," I pointed out. "Or, slightly less trivial, three numbers x, y, z satisfying the three equations

$$x + 2y = 3, \quad y + z = 2, \quad x - 2z = 1.$$

And yet no solutions exist." But he went back to differential equations.

Once he asked me about my explication of the ideas of curves, of ramification, and of dimension, which Birkhoff had mentioned to him. He listened with great interest and asked very good questions. When I saw him the next time he said, "I have thought over what you told me; and I see a similarity between your explications of geometric ideas and the explication of economic ideas in the theories of value, of capital, and of money."

He invited me to attend the charter meeting of the Economic Society to be held in Cleveland in connection with the Christmas meeting of the American Association of the Advancement of Science, where I had been invited to speak about the axiomatics of dimension. From there, Schumpeter returned to Germany via Japan.

9.

From Cleveland, I went to Chicago, where Eliakim Hasting Moore had invited me to spent New Year's Day 1931 with him. If there has been a father of American mathematics, then it certainly was Moore, the teacher of Birkhoff, Oswald Veblen, Robert Lee Moore, and most American mathematicians of their age who later became prominent. In 1930 Moore was old and unwell. He asked me about intuitionism, dimension theory, and metric geometry. He, too, was especially interested in metric betweenness. (Early in this century, as a student of Hilbert's, he had eliminated one of Hilbert's axioms about betweenness by proving it from the other assumptions.)

But Moore's great achievement was the foundation of General Analysis, based on the "fundamental principle of generalization by abstraction," which is also of some philosophical interest. "The existence of analogies between central features of various theories implies the existence of a general theory which underlies the particular theories and unifies them with repect to those central features." Today, mathematicians rarely speak about Moore, while carrying his principle to the limit (and sometimes beyond!). Moore gave me a copy of his *General Analysis*[5] with a dedication that I greatly treasure.

10.

From Cambridge, I repeatedly visited New York City, which made an overwhelming impression on me. On one occasion, I gave a lecture at Columbia University. A somber one-armed man attracted my attention; and when I asked someone about him I learned that his name was Emil Post. I told Post how much I admired his Thesis and I considered it a classic in the field of mathematical logic. Apparently no one had ever talked to him in these terms before and he seemed very gratified.

Early in February I arrived at the Rice Institute in Houston. I had not seen Gödel since June, so I learned of his epoch making logical discovery from a letter of Nöbeling while I was lecturing in Houston. I at once presented the ideas to a greatly interested, if somewhat bewildered group at the Rice Institute, where Charles W. Morris was then teaching as a young lecturer.

APPENDIX

Sheffer's results can of course also be expressed in the symbolism of the theory of functions.

Instead of Apq, Bpq, Cpq we may write $A(p,q)$, $B(p,q)$, $C(p,q)$. Then (α) and (β) read

(α') $B(p,q) = A(A(p,q),A(p,q))$,

(β') $C(p,q) = A(A(p,q),q)$.

Twenty years after my talks with Sheffer I noticed a difference between (α') and (β') that is mathematically, if not logically, highly significant. In the theory of functions, by substituting the pair of functions (G,H) into the function F one obtains a function denoted by $F(G,H)$. If its value for (x,y) is denoted by $(F(G,H))(x,y)$ then, by definition

$$(F(G,H))(x,y) = F(G(x,y),H(x,y)).$$

Substituting *into* a function and substituting an expression *for* a letter in logic (see VIII,3) thus have quite different meanings. Using *substitution into* one can write the right side of (α') as $(A(A,A))(p,q)$ and hence the whole of (α') in the form

(α^*) $B = A(A,A)$.

This formula expresses B in terms of A. In order to do something similar for (β'), however, one must introduce a particle *the first of*, indicated, say, by I, so that $I(p,q)$ means *the first of* p,q, that is, p; in a formula, $I(p,q) = p$. Then one can write (β') in the form

$$C(p,q) = A(A(p,q), I(p,q)) = (A(A,I))(p,q)$$

and hence

(β^*) $C = A(A,I)$.

But (β^*) expresses C in terms of A and I, and not in terms of A alone! In fact, I have proved that only 4 of the 16 binary particles of the 2-valued logic *can* (in this strict sense!) be expressed in terms of A alone, namely, A itself, B, tautology, and contradiction (in the sense of VIII,3). The 12 other particles are connected with A only in the looser sense that their values for (p,q) can be expressed in terms of A, p and q.

Notes

[1] Except for minor differences in the note in VIII.

[2] Another set satisfying the axioms I-IV consists of the numbers 1, 2, 3, 4 (as points) and the pairs { 1, 2 } , { 1, 3 } , { 1, 4 } , { 2, 3 } , { 2, 4 } , { 3, 4 } (as lines). An isomorphism with the tetrahedral model can be established by associating the numbers 1, 2, 3, 4 to the vertices and the pairs { 1, 2 } , { 1, 3} , . . . { 3, 4} to the edges joining the points corresponding to 1 and 2, 1 and 3, . . . 3 and 4, respectively.

[3] Percy Bridgman, *The Logic of Modern Physics*, New York, Macmillan Co, 1927.

[4] Josef Schumpeter, *Das Wesen und der Hauptinhalt der theoretischen Nationalökonomie*, Leipzig, Duncker & Humblot, 1908.

[5] [Eliakim Hastings Moore, *General Analysis*, (Philadelphia, The American Philosophical Society, 1935). Since this book was actually published after Moore's death, Moore must have given Menger one of his memoirs or perhaps his lecture notes which served as the basis for the book. L. G.]

XIV. DISCUSSIONS IN THE CIRCLE
1931-34

When I returned to Vienna in the fall of 1931 Feigl had decided to stay on for good in America while Carnap had left for Prague but visited the Circle from time to time, always provoking particularly lively discussions. He no longer talked about 'the' language and in fact, to the displeasure of Waismann and to some extent of Schlick, *used* several languages; but I continued belaboring the point until (I believe after his second visit) he also gave up references to 'the logic' despite the strong protests of Waismann and Schlick.

Carnap and Neurath had begun to develop their ideas on what they called *protocol* sentences, such as propositions of the following scheme:

> X (name of an observer) *observed on* (date and hour) *at* (name of location) *on* (name of the instrument) *the coincidence of the pointer with* (mark, say, 1.7).

Protocol sentences were presented as the basis of our knowledge — a claim that Schlick contested from the beginning. He called such sentences hypothetical and subject to later revision. According to Schlick, the bases of knowledge were what he called *Konstatierungen* — a word without a precise equivalent in English, which however can be easily exemplified by the observer X in the preceding scheme saying at the time and place mentioned above:

Here and now, the pointer on the instrument coincides with the mark 1.7.

The ensuing discussion was among the many which I followed silently.

At that time, another issue came up about which Carnap and especially Neurath were very vociferous, and would remain so for the next years. In his book *Der logische Aufbau der Welt*,[1] Carnap had (even before coming to Vienna) begun to execute a program for "constituting the world out of immediate sense-data by the methods of modern logic." Schlick and the other members of the Circle greatly appreciated this synthesis of Mach's and Russell's ideas; in Hahn's opinion it was the very core of logical empiricism. In the course of the years, various modifications of the original plan had proved to be necessary, especially in connection with the Circle debates on verification and confirmation (most of which, to my great regret, took place before I joined or while I was absent.) But one tenet remained permanent: Psychological, sociological and economical entities could be and were to be constituted from sense-data by logical methods just like the entities of the physical world.

On the other hand, various scholars including philosophers, sociologists, and historians, had for a long time maintained that they were in possession of special insights into ideas and of special methods of obtaining knowledge — insights and methods unavailable to the sciences. In German, the polarization is expressed by the words *Geisteswissenschaften* versus *Naturwissenschaften* (roughly, human vs. natural sciences).

The oldest and most common claim of this type, the pretension to an intuition of essences, had been treated by Schlick as early as 1913 in an excellent paper critical of Bergson and Husserl. "Is there intuitive knowledge"?[2] In the period between the two world wars, the claims became more numerous and more specific. They included the understanding of the meaning of history, empathy into the spirit of nations, etc.

Neurath and Carnap denied the social sciences all such privileged methods. They only granted them concepts that were in principle obtainable by constitution in the last analysis from sense-data and propositions that were amenable to testing by verification and confirmation just like those of the natural sciences. They promoted the idea of a unified scence. Schlick was not as enthusiastic about the project as some members of the Circle including Hahn, but favorably inclined.

For my part, I found what Carnap said about special cases of methodological illusion or abuse to be fully convincing. Yet, apart from an instinctive aversion to monistic schemes of any kind, I feared that the idea of a unified science might possibly lead to the exclusion a *priori* of potentially valuable objects or methods of study. And there was another point. Mutual delimitations of branches of science are necessary for the organization of studies and research of schools and libraries, and for diverse commercial purposes. Such delimitations are historically conditioned if not altogether arbitrary. But they are of hardly any epistemological interest. Attempts to freeze such delimitations in rigid definitions are not only useless but likely to slow down normal development and to inhibit progress by limiting a priori the objects and/or the methods of research. Now the unification of science seemed to me to repeat this objectionable procedure on a higher level. Science is being delimited — from what? from technology? from art? It is doubtful that this can be achieved except by arbitrary dictates; and it is even more doubtful that anything would be gained if it could be achieved.

Many scholars used the supposed insights and methods that were allegedly unavailable to exact science for spurious derivations or justifications of their political views, most of which happened to be extremely rightist. Earlier in the century, such arguments had played some role in the promotion of a fascist mentality in the higher schools of Germany and Austria. But when Neurath tried to use the unity of science movement to brake that development, the situation in Central Europe had gone far beyond the point of intellectual argument.

Notes

[1] Rudolf Carnap, *Der logische Aufbau der Welt*, Berlin-Schlachtensee, Weltkreis-Verlag, 1928, xi, 290pp.

[2] Moritz Schlick, 'Gibt es intuitive Erkenntnis?', *Vierteljahresschrift für wissenschaftliche Philosophie und Soziologie* 37, 472-48, also translated *Philosophical Papers I, 1909-1922*, Vienna Circle Collection 11, edited by Henk L. Mulder and Barbara F. B. van de Velde-Schlick, Dordrecht and Boston, 1978.

XV. THE CIRCLE ON ETHICS

1.

Except for aesthetics, ethics was the part of traditional philosophy least discussed in the Circle.

In their readings of the *Tractatus* in 1925-27, the group cannot have missed the esoteric aphorisms about value on the last pages of Wittgenstein's book: the assertion that there is no value *inside* the world (Tr. 6.41), immediately followed by the curious remark "and if there were one (namely, a value *in* the world) it would be of no value"; moreover, since "what is of value" can only lie *outside* the world, "there cannot exist ethical propositions" (Tr. 6.42); and "It is clear that ethics cannot be expressed" (Tr. 6.421). But in the mid-1920's, Schlick does not seem to have paid much heed to this part of the *Tractatus*. For example, Tr. 6.521 claims that "the solution to the problem of (human) life is seen in the disappearance of the problem" and suggests that for this reason "men to whom after long perplexity the sense of life became clear were unable to say wherein this sense consisted." Yet in 1927, Schlick published a booklet about the sense of life (*Vom Sinn des Lebens*),[1] in which he mentioned neither Wittgenstein nor the *Tractatus*. Rather, in almost poetical style, he extolled youthfulness and play.

Even in 1930, when Schlick's epistemological views were profoundly influenced by the *Tractatus* and conversations with its author, he published a book, *Problems of Ethics*,[2] in which, without any mention of Wittgenstein's ideas, he proposed to develop ethics in a strictly objective spirit. "Ethics seeks nothing but knowledge." That search, he continued, rules out the encroachment of wishes, hopes, and fears upon objectivity; it is incompatible with the positing of commands as well as with an absolute justification of norms. Only relative justifications are possible, basing rules on higher rules; but that procedure ends in ultimate rules or in a highest norm, which cannot be justified in the same way. In particular, Schlick criticized Kant's attempt to base a morality of duty on an absolute, formal command, the categorical imperative, "Always act according to that maxim of which you can wish that it should become a general law." Schlick emphasized that this absolute command had no commander — a remark echoed by several members of the Circle — and he likened it to an absolute uncle without nephews and nieces. According to Schlick, the object of ethics is rather the *explanation of human behavior* — essentially a psychological task which he carried out in the spirit of hedonism. Decisions of the individual are determined by the motive of greatest pleasure and least pain — altruistic behavior and even sacrificial suffering being among the possible sources of pleasure. Kant's ethic of duty Schlick attempted to replace with the ideal that individuals should desire what society demands — an ideal that seems to him to presuppose gradual changes both in the attitudes of individuals and in the claims of society. *Problems of Ethics* is a book breathing its author's spirit of gentleness and kindness, while the influence of analytic thinking is merely implicit in its avoidance of some of the worst traditional verbiage.

Carnap's early ethical statements were also quite restrained. In *The Logical Structure of the World* [3] he quoted with appreciation Ostwald's derivation of values from energetics and observed that the construction of values from certain value experiences is in many ways analogous to the construction of physical things from perceptual experiences. In his *Pseudoproblems in Philosophy*,[4] ethical questions are not mentioned. It was in a brief paragraph of his paper 'The vanquishing of metaphysics through logical analysis of language' (*Erkenntnis*, v. 2, 1932)[5] that Carnap first denied that propositions can express value judgments. For the objective validity of a value can be neither empirically verified nor deduced from empirical sentences. If predicates such as 'good' and 'beautiful' are described by empirical characteristics, then sentences including those predicates are statements about empirical facts and not value judgments. If no such characteristics are given, then he called those sentences pseudopropositions. All in all, this remained Carnap's view on ethics during the years of Schlick's Vienna Circle.

Neurath, in 1931, turned to the integration of sociology into his unified science. Ethics, he claimed, after the elimination of all theological and metaphysical elements was reduced to the behavioristic study of certain modes of human behavior and of the commands of persons to other persons. While viewing with disfavor all studies of the motivation of individual actions, Neurath pointed out the possibility of a *felicitology* in the sense of behavioristic study of the degrees of happiness under various regulations of life.

In 1936, the year when with Schlick's death the Circle in Vienna came to an end, Victor Kraft began writing a book *Grundlagen einer wissenschaftlichen Wertlehre* (*Foundations of a Scientific Theory of Value*),[6] which appeared in 1937. In it, Kraft distinguished two elements of a value judgment: the valuation proper and the content or object of the valuation. That content may be related to the contents of other value judgments, for example by subsumption. Relations of this kind between the objects account for logical relations between value judgments. In this way Kraft explained, for example, the deducibility of one value judgment from another.

2.

While the political situation in Austria during the winter of 1933-34 made it extremely difficult to concentrate on pure mathematics, socio-political problems and questions of ethics imposed themselves on everyone almost everyday. In my desire for a consistent, comprehensive world view I asked myself whether some answers might not come through exact thought.

I appreciated Carnap's demystified form of Wittgenstein's removal of traditional ethics from the realm of the cognitive, even though it certainly did not come to me in any way as a revelation. I had already been convinced by Hume and Poincaré that propositions about *what ought to be* could not be inferred from propositions about *what is*; and from this conviction, combined with an empiricist attitude, there is only a small step to Carnap's epistemological view of ethics. But what troubled me was the negativity of this view. Might not exact thinking, despite everything, yield *positive* ideas connected with ethical problems or at least be somehow applicable to them? I searched the literature of ethics.

After rereading Brentano's tracts expounding 'ethical intuitionism' — only to find the empty phrases that I remembered (See III.3) — I looked again into Kant's development of the categorical imperative. The very formulation of that principle is not the same in all of Kant's writings. He commands us to adopt maxims *that can be general laws* in one of his books, and maxims *that we can wish to be general laws* in another. Kant was of course entitled to change his opinion; but the difference just quoted — roughly comparable to that between formalism and intuitionism in mathematics — was not clearly brought out, let alone adequately emphasized. What troubled me even more, however, was a serious shortcoming from a strictly pragmatic point of view of the imperative in either form: the impossiblity of ascertaining what specific actions the imperative commanded in specific situations. For an individual cannot know which general rules are possible nor even state which rules he finds desirable unless he is aware of all, or at least the most important, consequences of the general rules in question. Most situations can be regulated in several ways: various laws are possible; more than one may even be desirable to the same person at the same time. In such situations the categorical imperative is of no help to an individual in making decisions. Moreover, the principle fails to give any hint as to what should happen if various persons make unlike and perhaps incompatible choices in their bona fide compliance with the imperative; nor does it indicate how one should react to intentional violations of it by others. Schlick had criticized Kant's imperative as a command without a commander; to me even more disturbing was the fact that the imperative did not permit the inference of specific commands.

In the more recent literature at that time, especially the German literature, I found books of four types: *sermons* which openly or in a veiled form preached a human behavior conforming to what, in the last analysis, proved to be the taste of the respective writers — sermons which, undoubtedly, were of interest to the biographers of those authors but not in theoretical ethics; *metaphysical* writings, such as Nikolai Hartmann's and Max Scheler's in whose intuitive vision of absolute values I failed to find, hard as I tried, positive ideas of the kind I was seeking; lengthy *methodological* discussions of the object and the procedures of ethics, many of them combined with attempts to define good and evil; and finally *empirical* studies. Some books of the last type revealed an immense wealth of historical, ethnographic, sociological, psychological and biological examples supplying illustrative material for an exact treatment of human relationships and organizations, but they gave nothing more than examples.

What I had been looking for, however, was a *theory* of ethics — an application of exact thinking that would bear to traditional ethics a relation somewhat comparable to that of mathematical to traditional logic. Even after all systems of morality, codes of norms and value judgments have been expelled from the cognitive domain and relegated to the realms of feelings or wishes, does there not remain, I asked myself, *a residue that lends itself to observation as well as to theoretical treatment* — perhaps *from an extensional point of view?* What indeed remains is *the group of adherents to the systems, to the codes and to the value judgments as well as the relations between such groups.* There existed, of course, descriptions of the groups of people obeying specific ethical norms as well as histories of the relationships between such groups. Descriptions and, even more, classifications of social (if not particularly ethical) processes, groups and relations also constituted the content of the work of L. von Wiese and his school. But even those studies, though dealing with observable social phenomena and hence more substantial than the vague statements of many other sociologists, had little to do with exact thinking. A *general theory* of relations between individuals and individuals, between individuals and groups, and between groups and groups resulting from diverse characteristics and attitudes of human beings and from their diverse demands on others — such a theory seemed to be nonexistent in the literature. I fully realized that ethics thus externalized would be regarded by most philosophers as quite superficial. On the other hand, it would lend itself to sound applications of the logic and mathematics of classes and relations and thus might bring about what I had fervently hoped for: a positive contribution, however modest, by exact thinking to questions of ethics. So during the winter of 1933-34, I tried to develop such a theory in a short booklet entitled *Moral, Wille und Weltgestaltung, Grundlegung zur Logik der Sitten,*[7] which I shall now discuss in some detail.

3.

The booklet begins with a letter to a friend in which I list topics that he should *not* expect me to treat: the *essence* of good and evil; the *foundations* of ethical norms; any single (proposed or commended) ethical system or in fact *any value judgments* whatever. Nor do I try to delimit ethics as a branch of knowledge or discuss its object and its methods. In short, the list of topics that I propose to omit covers almost all that traditional books on ethics discuss.

The second chapter consists of five epistemological notes on ethical subjects. I describe the genesis of an individual's concept of the good and his use of the word 'ought' as behaviorists and linguistic philosophers do today, of course in a much more elaborate form. I then point out that the various classical foundations fail to provide individuals with guidance in specific situations — a point illustrated in the preceding section of the present chapter with regard to the categorical imperative. In particular, I satirize the treatment of the question, "What is good?" by ethical intuitionists, more specifically, by Brentano (though without mention of his name). I compare the answer, "Good is what is worthy of love or what rightly ought to be loved," supplemented by, "What is worthy of love is evident," to an equally informative reply to the question, "What medicines are beneficial?"; namely, the answer, "Beneficial medicines are those worthy of ingestion or those which it is appropriate to use" supplemented by "What is appropiate to use is learned by experience." Experience is of course capable of a clearer and more specific description than is intuition. That, however, is not the point; the point here is that in both cases all that is interpolated between the questions (as to what is good and what is beneficial) and the answers is nothing but verbiage ripe for Occam's razor. Yet I carefully avoid characterizing such answers as *meaningless*. While the Circle in 1934 was still indulging in a great deal of ill-defined talk about meaningful and meaningless, I strictly observed my maxim, *Assert or don't assert or say that you will not assert*. In particular, I explicitly stated in the booklet that I would not call any sentence about good or evil *meaningless*. In fact, I compared anyone using this pejorative term to a man who sees his mortal enemy inescapably mired in a swamp and, instead of going on his way, rushes toward him with a drawn dagger.

The third chaper is devoted to a dialogue with my friend, who compares a value free ethics without any concepts of good and evil to a zoology ignoring animals, and doubts that such studies can be assigned to ethics at all. I express indifference to the classification, delimitation or definition of *any* branch of knowledge because of the fluidity of the topics of all of them, which I illustrate by examples from geometry. Since my friend seems to misunderstand me as suggesting that ethical norms are comparable to 'Euclidean truths,' I explain the existence of various geometries, incompatible with Euclid's and with each other. No one believing in *one* morality can, therefore, assert an analogy between ethics and geometry. Moreover, for my part, I do not believe that analogy to be far-reaching even though I am convinced of the existence of diverse moralities as well as diverse geometries. What then, my friend asks at the central point of the dialogue, is it that motivates a person to accept a definite system of norms? I tell him that the actual modelling of an individual's behavior according to some particular code or to some specific morality is ultimately based on a decision by that person — an answer that points in the direction of the ethical view propounded in the later 1930's by the French existentialists. (But while Sartre tried to buttress this view with a metaphysics of being and nothing, à la Heidegger, I based it simply on a logical analysis of ethical material.) In the case of a religious believer who derives moral precepts from superhuman commands, the decision lies in the acceptance of the underlying religious dogma. But my friend asks, does the thesis about decision in conjunction with the negativistic views expressed in my epistemological notes permit me to say anything *positive*?

The fourth chapter is devoted to positive statements of extensional ethics on groups of people and relations resulting from their moral attitudes. To each norm there correspond three classes: all those approving of the norm; those indifferent to it; and those disapproving of it. Special attention is given to relations resulting from the interplay of people's behavior and their demands concerning the behavior of others, and the book develops models of the situation, if only very simple models — in 1934, models of social phenomena had not yet come into fashion. One such model is based on the assumption that a population is divided in two ways: 1) into two *fundamental groups*: the group of the individuals possessing a certain characteristic or adopting a certain norm and the group without the characteristic or adoption of the norm, and (2) into four disjoint groups in which the interrelations reflect the willingness or unwillingness of individuals to associate with others on the mere basis of their possessing or not possessing the said characteristic or adoption of the said norm. This, incidentally, is the point where the preceding model greatly simplifies social reality. There, most forms of voluntary association are based on the individual's possession or non-possession of *several* characteristics (adoption or rejection of *several* norms). One of the simplest theorems concerning this model (that is, one of the immediate consequences of the assumptions) is the enumeration and description of all possible types of cohesive groups.

The fifth chapter concludes the book with a letter reviewing previously discussed matter from various angles.

Imagination is invoked in a large-scale thought experiment. On an autonomous island, which is in the course of being settled, the entire population is gathering (somewhat in the style of an ancient Greek assembly in the agora or a New England townhall meeting) in order to regulate social coexistence, either *all* aspects of coexistence without any preexisting constitution, legal code or common faith; or *some* aspects under certain assumptions about the others. Problems will be raised in the assembly for some of which more than one regulation (including in some cases nonregulation) will be proposed. After debating their motivations the individuals express their choices or preferences. But the settlers recognize the ultimately noncognitive character of these decisions as well as the fact that even obedience to the categorical imperative is neither necessary nor sufficient to achieve universal harmony. Applying this negativistic part of ethical criticism to the experiment, they decide *to limit each regulation to the group of those who favor it*. While the positive part of extensional treatment of ethics and of value judgments in general suggests to them the consititution of such groups subject to two

conditions 1) *Internal cohesiveness of each group*, though not necessarily homogeneity or consensus of its members, 2) *compatibility
of any two of the groups* to such an extent as to render peaceful
coexistence possible, though not necessarily to the point of coopera-
tion. To achieve internal cohesiveness of groups use is made of the
insight that likeness and even consensus of individuals is neither a
necessary nor a sufficient condition for their compatibility. Simple
examples of cohesive though nonhomogeneous groups include
groups of tolerant individuals of various characteristics and/or attitudes as well as Nietzsche groups, especially centered groups.
Among the problems that beset the constitution of coexisting groups
with incompatible codes the following four stand out (in order of
increasing difficulty, a) the validity of unlike codes in the same
territory; β) the desire of more than one group for the exclusive
possession of one and the same object as the basis of their formation, for example of the same tract of land; γ) the decision of one
group to influence the decisions of the members of another group —
a decision that may even be incorporated in a code of norms as was
the aggressive proselytizing of the early Muslims; δ) decisions of, as
it were, a higher order; that is decisions of a group concerning the
very system of grouping, such as the claim that a code be universal
and hence only a single group exist, as in the program of com-
munism, or that certain groups be exterminated in a Nazi-type
genocide. Some of these can be solved, others seem to be fun-
damentally insoluble problems.

   In social reality, however, a plurality of regulations for ethical and
social problems, each regulation confined to its adherents, are rarely
even *aimed* at. Most populations, be they large or small, look for
general solutions and uniform regulations encompassing all members. The principal methods for achieving this aim are dictates and

compromises. Dictates issue from a single dictator or a few persons (in fascism) or from a particular class distinguished by certain characteristic such as ancestry or achievement (in feudalism or élitism) or from a plurality or the majority of adherents (in democracy); they may be opposed by a significant part of the population, even by a large majority. Compromises between unlike codes are established by modifying conflicting norms toward an intermediate formulation and/or by selecting some regulations from each of the codes and combining them in an eclectic system of norms; they introduce adulterating modifications into a plan. For better or for worse, noncooperation of a significant part of the population as well as modifications of a universal plan may be responsible for the eventual failure of a principle. Hence both dictates and compromises often make it impossible to arrive at a clear judgment concerning (past or present) socio-ethical projects and regulations, especially concerning their consequences. What is more important, neither system allows a full and free competition of several ideas; and many plans can never materialize at all. In every form of social organization, there will of course be socio-ethical plans or combinations of projects that cannot be realized on a voluntary basis; for this would require more participants than they have adherents. This is a simple consequence of human inventiveness in planning. Under uniform regulations, however, even many projects and combinations of projects with sufficiently many adherents for their realization cannot come to fruition.

Since the book aims at strict objectivity, it eschews value judgments even with regard to the highest level of ethics; the words 'good' and 'evil' as well as all related terms of evaluation are meticulously withheld not only from norms and codes but also from forms of organization, such as the constitution of compatible cohesive groups. Only in the closing section of the book is it mentioned that such a constitution, wherever it is feasible, would correspond to the author's personal taste; and strong emphasis is laid on the purely subjective nature of this remark. But in perfect harmony with strict objectivity, *the extensional logic of ethics points out the possibility of forms of organizations that are traditionally overlooked or neglected.* Thus what the book suggests is 1) the study of general conditions for the constitution of cohesive mutually compatible voluntary groups; 2) in specific cases, the search, in the direction indicated, for alternatives to dictates and compromises.

While both the critical and the positive parts of the study are strongly opposed to the categorical imperative another less known dictum of Kant's seems to be in better harmony with voluntary groups. In one of his explanations of morality, Kant wrote that "a legislation ought to be found in every reasonable person and must be able to arise from his will." If one realizes the multiplicity of possible regulations of a socio-ethical problem (which seems to have escaped Kant completely), then the search suggested in my book almost appears as a quest for the realization of that dictum of Kant's.

4.

A reaction reached me, indirectly, from Kraus at the German University of Prague — the principal apostle of Brentano's ethical doctrine. Probably incensed by my somewhat irreverent quotations from Brentano's writings Kraus wrote an irate letter to a Viennese cousin of his, Mrs. Ella Dub, widow of a prominent economics journalist. Kraus knew that Mrs. Dub was acquainted with me. In violent terms, Kraus denouced my book as incorrect and immoral. Mrs. Dub, who had been following my career with a somewhat motherly interest, was greatly alarmed, spoke to me at once and then discussed my book and Kraus' letter with Schlick, whom she knew well. But, as Schlick told me the next time we met, despite sincere efforts he succeeded only partly in reassuring her. So Mrs. Dub decided to consult her most revered friend and supreme authority, who was expected to deliver a lecture in Vienna — Max Planck. She gave a lavish party in honor of the Plancks and found an opportunity to hand to him the material that she had been discussing with Schlick. Naturally Professor Planck did not mention the matter in a conversation he had with me during the party. By the end of his stay in Vienna, however, he seemed to have looked though Mrs. Dub's dossier rather carefully. As a result, Planck completely reassured her, as she told me with every sign of relief, that he considered my approach to ethics to be neither incorrect nor immoral. The only criticism of the book that Planck expressed was that — as incidentally is also pointed out in so many words in its very pages — it of course by no means covers *all* ethical problems.

Clearly, such a study does not aim at creating or expanding reducing or dissolving groups associated with decisions, — aims that motivate the founders of religions, preachers and reformers, or Nietzsche and Tolstoy. For, however strongly the theses of all these men may have conflicted basically they all had one trait in common: they proclaimed value judgments in one way or another, openly or in hidden form, which the study here outlined does not. And in contrast to many writings that merely pay lip service to the idea of evaluation-free cognition, this work actually is completely free from value judgements. What can be reaped from the path outlined should, therefore, in a proper perspective, not be compared with the effects of apologies for egoism or glorifications of altruism, of a commendation of

hedonism or a gospel of asceticism, designs for socialism or proposals of individualism — proclamations of value statements which,
disseminated with eloquence, have indeed deeply influenced some
individuals in establishing their personal aims. It should rather be
compared with the fruits, if any, of traditional ethical *theory*: of the
various attempts to define good and evil or of the proofs of their
indefinablility or of the claim to intuit absolute values or of ethical
agnosticism.

Notes

[1]Moritz Schlick, 'Vom Sinn des Lebens', *Symposion*, (1927) 331-354; also published
as no. 6 of the Sonderdrucke des Symposion (Eng. transl. in *Philosophical papers II,
1925-1936*, Vienna Circle Collection 11, II edited by Henk L. Mulder and Barbara F. B.
van de Velde-Schlick, Dordrecht and Boston; 1979).

[2]Moritz Schlick, *Fragen der Ethik, Schriften zur wissenschaftlichen Weltauffassung IV.*
edited by Ph. Frank and M. Schlick, Vienna, 1932, *Problems of Ethics*, New York,
transl. by David Rynin, 1932 (further English editions 1961,1962).

[3]Rudolf Carnap, *Der logische Aufbau der Welt.* (Berlin-Schlachtensee,: Weltkreis-
Verlag, 1928. xi, 290pp; *The Logical Structure of the World, Pseudoproblems in
Philosophy*, trans Rolf A. George, Berkeley and Los Angeles, University of California
Press, 1967.

[4]Rudolf Carnap, *Scheinprobleme in der Philosophie: Das Fremdpsychische und der
Realismusstreit*, Berlin-Schlachtensee: Weltkreis-Verlag, 1928, 46pp (1967) see previous note.

[5]Rudolf Carnap,'Überwindung der Metaphysik durch logische Analyse der Sprache',
Erkenntnis Leipzig, Bd. 2, H. 4 (1932), 219-241.

[6]Victor Kraft, *Die Grundlagen einer wissenschaftlichen Wertlehre*, Schriften zur wissenschaftlichen Weltauffassung, Vienna XI. ed. P. Frank and M. Schlick, 1937; reprinted Vienna Circle Collection 15.

[7] Reissued in English as *Morality, Decision and Social Organization Toward a Logic of Ethics*, Vienna Circle Collection 6, Dordrecht and Boston, D. Reidel Publishing Co., 1974.

XVI. MORITZ SCHLICK'S FINAL YEARS

1.

In 1933 the year of Hitler's coming to power in Germany there were periods, when life in Vienna was almost intolerable. The newspapers published extras around the clock and vendors ran shouting through the streets offering the latest editions. Groups of young people, many wearing swastikas, marched along the sidewalks singing Nazi songs. Now and then, members of one of the rival paramilitary groups paraded through the wider avenues. I found it almost impossible to concentrate and rushed out hourly to buy the latest extra. On one of these days, I met Dr. and Mrs. Schlick in a street car. "It is impossible to concentrate," the professor said, "I read extras from morning to night."

Schlick's position at the University became precarious. He was not and as far as I knew, never had been politically active. His political views can probably best be described as those of a British-style liberal. But this was far from satisfactory to the nationalistic professors and students. Moreover, in all administrative and academic matters, especially with regard to proposals for new appointments, Schlick was always objective, unprejudiced and impartial — qualities that failed to ingratiate him with the majority of his colleagues at that time. And that majority had more specific objections to his activity: his insistence on keeping Waismann (who was Jewish) as his assistant; his supposed friendship with the radical Neurath; his Kreis (the Vienna Circle!) which, notwithstanding the strictly apolitical way in which Schlick conducted it, was regarded as a sort of secret conspiracy; and finally his sponsorship of the Ernst Mach Society, which was almost treated as an open rebellion (it would be dissolved by the Dollfuss regime in the wake of the first civil war in 1934).[1]

To some extent all of us were, of course, affected by these conditions making it difficult even for the most faithful friends of Austria to feel about the country as they had before. It was sad also to see Schlick's quiet serenity slowly disappear. In one of my conversations with him during that terrible period he said that in his opinion Hitler meant the *Untergang des deutschen Volkes* — the decline and fall (more precisely, the perdition) of the German people; and that all should join in a supreme effort to avert this imminent calamity. He also mentioned to me his intention to write a letter to Cardinal Innitzer, the archbishop of Vienna, whom many of us had met when he was professor of theology at the University. We had admired him for restoring that institution for the year of his rectorship to a place of teaching and learning: he energetically suppressed riots of Nazi students that constantly plagued the university before and after that year and forced most other rectors to keep the institution closed for long periods. In his letter to the cardinal, while mentioning his unrelatedness to Catholicism, Schlick planned to offer his support of any action that might stem the flood.[2] (Whether Schlick ever dispatched that letter I don't know).

In 1933-34, the university was closed for extended periods of time. Both Schlick's Circle and my Colloquium, however, met regularly though Schlick, Hahn, and I, being the only members with keys

to the deserted buildings, had to let the others in. Upon entering one
had the feeling of having reached a quiet oasis.

Austria's chancellor Dollfuss, who ruled without Parliament, not
only wanted to rid the country of Nazis but also tried, probably
abetted by Mussolini, to destroy their only absolutely implacable
Austrian opponents, the Social Democrats. He subjected the latter to
provocations culminating in the suppression of their widely read daily
newspaper. At the beginning of 1934 an explosion was clearly imminent. In February, Dollfuss under some pretext attacked the main
city-owned buildings (chiefly the *Karl Marx Hof*) of the organized
Viennese laborers with heavy artillery. The ensuing civil war ended
with the destruction of the Social Democrat forces. Hitler's Austrian
followers, greatly strengthened by these events, made ever increasing demands on the government. In July 1934, in a surprise attack
on the chancellery, a group of Nazis assassinated Dollfuss thereby
starting a short civil war. But the new chancellor, K. von Schuschnigg, succeeded in suppressing that uprising; and he continued his
own brand of fascist regime until Hitler occupied Austria in 1938.

The day before Dollfuss was murdered, Schlick and the few left
in the Circle were deeply shocked by the totally unexpected death of
Hans Hahn after a short illness, Neurath, who happened to be
abroad in February, never returned to Austria.

2.

There was in Vienna in the mid 1930's one respected politician who
kept out of all the conflicts — the *Bundespräsident*, President Miklas
of Austria. I had never seen him in person, but in the spring of 1936 I
received from his office an invitation to attend the opening of some
exhibition in a building of the former imperial court. I arrived a little

late and found that the room was so packed with guests that it was almost impossible to move. The first familiar person whom I saw was Schlick standing not far from the entrance. We were talking when a passage opened in the crowd near us. Through it, the President, having completed the opening ceremony, was leaving with his party. To my surprise, one of the men behind the President waved rather intimately to Schlick, and the latter responded in the same way. "You have friends in the government?" I asked teasingly. But Schlick's expression changed to one of utmost seriousness and he said in a grave tone, "That is not a friend. That is a security man who used to be my body-guard." I must have looked totally bewildered so that Schlick explained further. "For some time now I have been threatened by an insane person who is in and out of mental institutions; and the man behind the president used to be assigned to my protection." "So you are no longer threatened," I said. Schlick sighed. "Until quite recently the fellow had been interned," he said. "But just three days ago he was released again; and yesterday I had another threatening telephone call from him. Yet, for all his threats, he has never actually harmed me. So I don't dare complain to the police again." And, as though it had happened only yesterday, I remember how Schlick added with a forced smile, "I fear they begin to think it is I who is mad," and he changed the topic. But I felt that he actually lived in great fear and must have done so for a number of years.

A few weeks later, Vienna was stunned by the news that, on the steps to the auditorium in the university building, a paranoic former student had shot and killed Professor Schlick.

The grief of the philosopher's numerous friends and students was deep . Since Schlick had held Austria's most prestigious chair in philosophy all newspapers published long, if uncordial obituaries. They also mentioned facts about the assassin. He was a disgruntled former student of philosophy who had the paranoic idea that Schlick frustrated his attempts to find employment. But as soon as the first impact of the tragedy had subsided some newspapers close to the government and paramilitary organizations changed their tone to open hostility. They now wrote that while murder was always condemnable it was not altogether surprising that students of Schlick's corrosive philosophy (zersetzende Philosophie) would etc., etc. . . that the instruction in philosophy at Austrian universities had to be reformed and so forth.

It would be futile to speculate about what might have happened had the bullets missed the victim or had the madman been permanently interned and Schlick allowed to live out his life without fear. But there is little doubt that he would have deservedly attained an even higher stature as a philosopher in the English-speaking world and eventually in Germany than he has, since being cut off from his work and further philosophical development. And many more would have enjoyed his wisdom and kindness.

Notes

[1] Nationalistic students tried to discredit Schlick even because of his given name, Moritz, which many Austrians (perhaps because of its alliteration with Moses) regarded as Jewish even though the name clearly derived from the Latin *Mauritius*. Moreover, Schlick was named after a close relative of his mother, the once famous Ernst Moritz Arndt, one of the greatest heroes of Germany in the Napoleonic period. His books and poems inspired the liberation of the country from French occupation. In his writings, Arndt developed ideas about German national aims which some National Socialist theoreticians, knowingly or unknowingly, paralleled. To the end of his long life in 1861, he was called 'the most German of all Germans' (*der teutscheste aller Teutschen*, *teutsch* being a chauvinist version of *deutsch*, related to *teutonic*).

[2] Because of his activity as rector, the cardinal appeared to Schlick as a permanent

pillar against Nazism. But those who hoped that he would stand up to the Nazis after Hitler's occupation of Austria were disappointed.

MEMORIES OF KURT GÖDEL *

1.

Soon after assuming my position at the University of Vienna in the fall of 1927, I offered a quite well attended one-semester course which was on dimension theory. The name of one of the students who had enrolled was Kurt Gödel. He was a slim, unusually quiet young man. I do not recall speaking with him at that time.

Later I saw him again in the Schlick Kreis — the group gathering in fortnightly meetings around the philosopher Moritz Schlick which became known to the world as the Vienna Circle. I never heard Gödel speak in these meetings or participate in the discussions; but he evinced interest by slight motions of the head indicating agreement, skepticism or disagreement. I, too, rarely took part in these debates; but, at Schlick's request, I did report in two sessions about the contents of my paper 'Bemerkungen über Grundlagenfragen' (Remarks about the foundations of mathematics.)[1] The second of these reports, which gave rise to what Carnap later called the Principle of Tolerance, prompted some rather unfavorable immediate reactions;[2] but I noticed that Gödel greeted my presentation with vividly approving nods.

In the academic year 1928-29 at the request of some students, I began directing a *Mathematisches Kolloquium* — an informal evening gathering with those students, who were soon joined by foreign visitors. In these meetings, topics and results in their and my fields of interest were reported and discussed. We followed the unconstrained style of the Schlick Circle; but from the fall of 1929 on I kept a *Protokollbuch* — minutes of a sort — something which Schlick, as far as I know, unfortunately never did.

Towards the end of 1929 I also invited Gödel to the Colloquium and from then on he was a regular participant, not missing a single meeting when he was in Vienna and in good health. From the beginning he appeared to enjoy these gatherings and spoke even outside of them with members of the group, particularly with Georg Nöbeling, sometimes with Franz Alt and Olga Taussky, when she was in Vienna, and later on frequently with Abraham Wald and foreign visitors. He was a spirited participant in discussions on a large variety of topics. (See below, Sections 4, 5, 8). Orally, as well as in writing, he always expressed himself with the greatest precision and at the same time with the utmost brevity. In nonmathematical conversations he was very withdrawn.

In the summer of 1929, visting the mathematicians in Warsaw I became acquainted with Alfred Tarski, Alfred Lindenbaum and their students. Impressed by their logical results (mainly concerning the propositional calculus) as well as by their treatment of philosophical problems by means of logical techniques, I wanted to make these investigations known both to my Colloquium and to the Schlick Circle and therefore invited Tarski to give three lectures to the Colloquium — two of them on logic; to these, I also invited the Circle. After the lectures, Gödel asked me to arrange a meeting with Tarski, since he wanted to report to the visitor about his doctoral thesis establishing the completeness of first-order logic. Tarski showed great interest in the result.

2.

In the summer of 1930 I did not see Gödel — not even in September, when the Schlick Circle met with a few congenial philosophers in Königsberg, where at the time the German *Mathematiker-Vereinigung* were holding a meeting. There, in a discussion on the foundations of mathematics, Gödel mentioned for the first time the discoveries he had made during that summer.[3]

I was absent, because I was sailing to America to lecture in several places. Nöbeling had agreed to chair the Colloquium during the academic year of my absence and to send me reports. I received them regularly. But early in February, soon after my arrival at the Rice Institute in Houston, Texas, a message reached me that left me thunderstruck. Nöbeling's letter began by recalling my often expressed admiration for analytic number theory — the idea being to prove theorems about the natural numbers 1, 2, 3, . . . by analytic means (that is, by means of the limit concept or even by profound continuity considerations) when no one had found a proof by induction — of course in the tacit expectation that sooner or later arithmetical proofs would be discovered. On January 21, 1931, Nöbeling continued, Gödel had presented to the Colloquium a paper entitled 'Über Vollständigkeit und Widerspruchsfreiheit' (On Completeness and Consistency),[4] in which he proved, *inter alia*, the existence of propositions about natural numbers that can neither be proved nor disproved by induction alone and thus, insofar as they are valid, *require* analytic methods of proof. One of these arithmetically undecidable arithmetic propositions is equivalent to the consistency of arithmetic, thereby shattering Hilbert's program for basing mathematics on proofs of its consistency. In my excitement about this news I interrupted my course with a report about Gödel's epoch-making discovery. Thus the mathematicians at Rice Institute were probably the first group in America to marvel at this turning-point of logic and mathematics.

3.

In the letter of congratulation that I wrote to Gödel expressing my admiration for his discovery, I mentioned a minor logical problem that had just occurred to me. The traditional introduction of logical operators (e.g. negation and implication, respectively indicated by ~ and ⊃) is based on the dichotomy of sentences into true and false ones. For example,

~ p is true if and only if p is false;

p ⊃ q is false if and only if p is true and q is false.

In the same way, tautologies are defined with respect to this bipartition of sentences, namely as combinations of sentences (e.g. in terms of ~ and ⊃) that are true no matter how the two truth-values are distributed among the component sentences. I asked myself whether a dichotomy of sentences into some that could be called 'true' and the others to be called 'false' was demonstrable from the axioms of the propositional calculus — formulas with *undefined* symbols satisfying certain conditions (such as the axioms of Frege or Hilbert or the Polish logician Lukasiewicz in terms of the undefined symbols ~ and ⊃). What I had in mind was, in other words, a *logic without truth and falsity* presented as a *pure algebra of logical operators* without the traditional semantic substructure. After a short time, Gödel sent me a mathematically rather simple proof of an affirmative answer to my question. He also presented this result to the Colloquium under the title 'Eine Eigenschaft der Realisierungen des Aussagenkalküls' (A property of the realizations of the proposi-tional calculus)[5] in June, 1931.

From an epistemological point of view, such a construction of logic may appear to be nothing more than a *tour de force* of little importance. Nevertheless, this result of Gödel's and the entire idea of logic without true and false seem to have received less attention than they perhaps deserve.

As described in my *Selected Papers* (p. 17f), a public lecture series, entitled *Krise und Neuaufbau in den exakten Wissenschaften* (Crisis and Reconstruction in the Exact Sciences), was offered at the University of Vienna, during the academic year 1931-32. The series began with empirical material and then proceeded to more and more abstract themes in a manner understandable to an educated general audience. The last lecture, devoted to logic, was assigned to me. I welcomed the opportunity to present the old logic from a more modern point of view and at the same time to present Gödel's great discovery not only in terms of its intrinsic significance, but also in historical perspective.[6] There was a further motivation in the fact that, in 1932, when Gödel's work was still almost completely unknown even among specialists, such a presentation had not yet been attempted. After the successful conclusion of the lecture series, it was published in a little book with the same title as the series.[7] Fa-vorable reviews arrived from all around the world — from America came one by the physicist and philosopher Percy Bridgman, who expressed special interest in my lecture on logic.[8]

4.

Gödel obviously continued to enjoy the meetings of the Colloquium, spoke with the other participants and was generous with advice in logical and mathematical questions. He always grasped problematic points quickly and his replies often opened new perspectives for the enquirer. He expressed all his insights as though they were matters of course, but often with a certain shyness and a charm that awoke warm and personal feelings for him in many a listener.

In the fall of 1931, I edited the first issues of the proceedings, the *Ergebnisse eines Mathematischen Kolloquiums* (subsequently for brevity, *Ergebnisse*) and asked Gödel and Nöbeling to be co-editors. Beginning with the reports about 1928-29 and 1929-30 which were excerpted from the minutes (the *Protokollbuch*), eight issues appered. The present section is devoted to a complete technical report on Gödel's logical contributions.

These began in Heft 2 (the second issue). The report he delivered in May, 1930, on his thesis is only briefly mentioned (2,17). But the issue contains a communication entitled 'Ein Spezialfall des Entscheidungsproblems der theoretischen Logik' (A special case of the decision problem for theoretical logic, 2,17f., *CW* pp 230-5), which Gödel wrote at my request in continuation of his thesis. In this note, he applied his method to the development of a procedure by which one can decide for each formula in a normal form whether it can be fulfilled. Issue 3 contains the proceedings of 1930-31 and on pp. 12f. and 20f. Gödel's two talks mentioned above in Sections 2 and 3.

Meanwhile, of course, our Colloquium continued with Gödel's active participation. Issue 4 of the *Ergebnisse* reports the results of the year 1931-32. It includes a paper 'Über Unabhängigkeitsbeweise im Aussagenkalkül' (On independence proofs in the propositional calculus, 4, 9f., *CW* pp. 268-71), in which Gödel, answering negatively a question of Hahn's, shows that not all independence proofs for the propositional calculus can be carried out by means of finite models. This also holds for Heyting's intuitionist propositional calculus.[9]

Gödel also deals with the propositional calculus in two further communications written for Issue 4. In one of them, 'Zum intuitionistischen Aussagenkalküls' (On the intuitionistic propositional calculus, 4, 40, *CW* pp. 222-225), he proves (again in response to Hahn) that there exists no model with only a finite number of truth values that precisely satisfies the intuitionist formulas (i.e. those and only those provable in Heyting's system); and he demonstrates at the same time that there exists a monotonically increasing sequence of systems between Heyting's and the classical propositional calculus. In the other communication, 'Zur Interpretation des intuitionistischen Aussagenkalküls' (On the interpretation of intuitionistic propositional calculus, 4, 39f., *CW* , pp. 300-3) Gödel formulates an interpretation in terms of the concepts of the ordinary propositional calculus and the notion 'p is demonstrable' (designated by Bp) assuming three axioms about B besides the axioms and rules of inference of the traditional propositional calculus. This system is equivalent to C. I. Lewis' system[10] for strict implication if Bp is translated into 'p is necessary' (designated by Np) and Lewis' system is supplemented by Becker's axiom: Np strictly implies NNp.

In June, 1932, in the last meeting of the academic year 1931-32 (which will be mentioned again in Section 7), Gödel presented a particularly important paper 'Zur intuitionistischen Arithmetik und Zahlentheorie' (On intuitionistic arithmetic and number theory, 4, 34-38, *CW* pp. 286-95). Heyting's propositional calculus obviously appears as a proper subsystem of the classical calculus if one replaces the intuitionistic notion of absurdity ($\neg$) by the classical negation ($\sim$). But also conversely, by a suitable interpretation of the negation operator, one can obtain the entire classical propositional calculus as a proper subsystem of Heyting's. In his Colloquium paper, Gödel proved a similar assertion for the whole of arithmetic and number theory (based on Herbrand's axioms). By a suitable interpretation of classical concepts in terms of intuitionistic ones, all theorems provable from the classical axioms are also valid in intuitionistic arithmetic and number theory, which hence are only apparently more limited than the traditional theories. "The reason for this," he writes, "lies in the fact that the intuitionist prohibition against negating universal propositions and against asserting purely existential propositions is cancelled in its effect by the fact that the absurdity predicate is applicable to universal propositions, which leads formally to exactly the same propositions as those asserted in classical mathematics."

5.

Gödel also participated actively in the geometrical investigations of the Colloquium, and this section contains a (perhaps pedantically accurate) survey of his contributions in that area. Non-mathematicians may skip this section.

In an interesting short paper (4,17f., *CW* pp. 278-81), Gödel deals with the metric concept of betweenness and its axiomatization. In a general metric space, if xy designates the distance between the points x and y, I had introduced betweenness by defining[11] that r lies between p and q if

$$p \neq q \neq r \text{ (and) } pq + qr = pr.$$

Wald had characterized this betweenness relation as a ternary relation in metric spaces having five properties.[12] Gödel now associated to every ordered triple (a,b,c) of points of the metric space a point of the 3-dimensional Eucidean space, R_3, namely the point (x,y,z) such that

(+) $x = ab,$ $y = bc,$ $z = ac.$

Because distances are never negative, M^3 (the set of all ordered point triples of M) is thus mapped into the first octant of R_3 — and, since M satisfies the triangle inequality, into the set W of all points (x,y,z) of that octant satisfying

(*) $(x + y - z)(x - y + z)(-x + y + z) \geq 0,$ $x,y,z \geq 0.$

The point b lies between a and c if and only if the point (x,y,z) corresponding to (a,b,c) belongs to the intersection of W with the plane $x + y = z$. Gödel then translated Wald's five properties of betweenness into five properties of subsets of R_3 that characterize[13] the set W. He closed his talk with the remark that his result suggested a study of the mapping (+) of M^3 into R_3 for metric spaces in general. This mapping has always appeared to me to be a sort of counterpart of Gödel's famous mapping of logical into arithmetical entities. But as far as I know, Gödel's interesting suggestion (which, by the way, can be extended to mappings of M^4, M^5, . . . into higher-dimensional Euclidean spaces) has not been carried out.

Gödel was also interested in the algebra of geometry, today known as the lattice-theoretical approach to geometry, which I had initiated in 1927,[14] and which was further developed in the Colloquium by Gustav Bergmann, Alt and Otto Schreiber.[15] After Schreiber reported his results, Gödel (4,34, *CW* pp.280-1) suggested investigating the set of sentences of that algebra that do not contain existential quantifiers in the prefixes of their normal forms. (The notions of *point* and *straight line*, which are defined using such prefixes, are not definable in this narrower set of sentences!)

Further, Gödel did not consider it beneath him to prove, in February, 1932, in reply to a problem raised in December, 1931, by Laura Klanfer (4,10), that every non-planar quadruple of R_3 is congruent with four points on a sphere for which the distance between any two points is defined as the shorter of the two great circles connecting them (4,16, CW pp. 276-9).

Finally, Gödel became interested in the direct metric definition of curvature $\kappa(p)$ of an arc A at a point p in a general metric space. By way of establishing a coordinate-free differential geometry,[16] I had defined $\kappa(p)$ as the limit of the reciprocal radii of the circumference[17] of the point triples (q,r,s) on A near p; and in the Colloquium, Alt strengthened this definition by considering just triples (p,q,r). Gödel suggested to Alt that he restrict the study even further to triples on A for which q and r lie on different sides of p (4,4).

Issue 5 contains a short note in which Gödel shows, in modern terms, that every automorphism in the real projective plane which maps straight lines into straight lines is bicontinuous and hence a collineation (5,1 CW pp.302-3). Mention is also made of a discussion between Gödel, Wald, and myself on higher-dimensional coordinate-free differential geometry (5,25) but no details about Gödel are there given or are now present to my mind.

I do remember that in 1932 Gödel began to study general relativity theory and to look for integrals of the field equations — a topic on which he published profound results 17 years later.[18]

6.

In addition, Gödel in those years also studied a great deal of philosophy including post-Kantian German idealist metaphysics. One day he came to me with a book by Hegel — unfortunately I have forgotten which one — and showed me a passage that appeared to anticipate general relativity theory. Gödel was convinced — and I agreed — that this was fortuitous, since Hegel could not possibly have intended the passage in the sense of Einstein. Still, the co-incidence was really quite amazing.

But Gödel had already begun to concentrate on Leibniz, for whom he entertained a boundless admiration. He keenly desired to inspect Leibniz' unpublished manuscripts and not only out of histori-cal interest, although I could not imagine him at that stage learning anything from Leibniz that would be relevant for his own work.

In 1933 he already repeatedly stressed that *the right* (*die rech-ten*, sometimes he said *die richtigen*) *axioms of set theory had not yet been found.*

Gödel no longer appeared regularly as a member of the Schlick Circle. The 'manifesto', published by Neurath (co-signed by Hahn and Carnap) in 1929[19] while Schlick was in America, alienated Gödel from the Circle to the point that he came to the meetings less and less frequently. We usually went away together from any sessions that we both attended, since we had the first stretch of our ways home in common. After one session in which Schlick, Hahn, Neurath and Waismann had talked about language, but in which neither Gödel nor I had spoken a word, I said on the way home; "Today we have once again out-Wittgensteined these Wittgensteinians: we kept silent." "The more I think about language," Gödel replied, "the more it amazes me that people ever understand each other."

Around that time Hahn began to arrange for Gödel's dozentship (*Habilitation*) at the University of Vienna.

7.

One day at the end of June, 1932, Professor Oswald Veblen, whom I had come to know in Princeton, telephoned on his way through Vienna and I invited him to spend a few days with me. Veblen was very preoccupied with the organization of the Institute for Advanced Studies, in the process of foundation at that time. I called together a meeting of the Colloquium during his stay in Vienna. Since I thought Veblen might one day be able to help Gödel in America, I wanted to give the visitor an opportunity to hear Gödel lecture (although of course I could not foresee what an important role he would play in Gödel's life). I therefore asked the latter to give in that session a talk on a remarkable discovery he had mentioned to me shortly before: that all of classical arithmetic and number theory (suitably inter-preted) is derivable in intuitionist mathematics (See above, Section 5). Veblen was very impressed by Gödel and invited him to visit the new Institute. It was too late to make arrangements for the year 1932-33, but it was agreed that Gödel should spend the year 1933-34 in Princeton.

In 1934 Gödel returned from Princeton to Vienna in poor health. But at the very first Colloquium session of the academic year (on November 6, 1934) he was again present.

8.

After his visit to America, Gödel seemed to be rather more withdrawn than before. The meetings of the Colloquium, however, he attended regularly and spoke freely with the other participants, who now came from all over the world. Nöbeling having returned to Germany, Gödel most frequently talked with Wald and Tarski, the latter being a participant in the Colloquium with a fellowship during most of the academic year 1934-35. Gödel also had discussions with Carnap whenever the latter came from Prague to Vienna for short visits to the Schlick Circle.

During Gödel's absence, Wald had begun to work, *inter alia*, on a problem of mathematical economics, connected with Walras' famous production equations,[20] which connect the quantities and prices of products with those of the factors of production. The number of the equations is equal to the combined numbers of the products and the factors. On this ground economists naively assumed that, given the quantities of the factors and the (expected) prices of the products, the equations determined the quantities of the products and the prices of the factors. But no one seems to have seriously tried to solve the equations since such an attempt would have made it obvious that in general solutions do not exist.

The Viennese economist Karl Schlesinger had found a flaw in Walras' equations.[21] They were based on the tacit assumption that all available quantities of the factors are actually used in the production. Schlesinger modified the Walras equations (in the form simplified by Cassell) by making allowance for unused quantities of factors and suggested to Wald, whom I had recommended to him as a tutor, that the equations so corrected might have solutions. By ingenious methods, Wald proved not only the existence of solutions but also their uniqueness.[22]

Gödel, when informed about these investigations, evinced great interest and asked Wald to bring him up to date, the first meeting of the year 1934-35 was devoted to a second paper by Wald[23]. on the subject — more specifically, to an elaboration of the remarkable fact that assumptions of a marginalist nature (that the expected price of products decreases with increases of their quantity) are indispensable for Wald's solutions. In the discussion after Wald's paper, Gödel said:[24] "Every producer's demand depends also on his income, which in turn depends on production costs. One might formulate a system of equations accordingly and investigate its solvability." This remark — to my knowledge the only one ever published by Gödel in the area of economics — testifies to his familiarity with the subject.

In June, 1934,[25] Gödel presented a paper, 'Über die Länge von Beweisen' (On the length of proofs)[26] — a continuation of the last paragraph of his 1931 paper on decidability and consistency (discussed above in Section 2). There, beginning with the formal system consisting of Peano's axioms, the scheme of recursive definition and first-order predicate calculus, Gödel had produced successive extensions by introducing variables first for classes of numbers, then for classes of classes of numbers, and so on. He had stressed that the consistency of each preceding system is provable in the successive systems; furthermore, that at every level there exist undecidable sentences that become decidable at higher levels. In the 1934 paper he pointed out that in transition to logics of higher order, not only do previously unprovable propositions become provable, but many proofs already available become greatly abbreviated. This paper was to be Gödel's last Colloquium publication.

In the fall of 1935, Gödel again travelled to Princeton to spend the academic year there. But he became ill, interrupted his stay in America and returned to Vienna in December in quite a bad state of health and mind. When the Colloquium meetings started and I had still not heard from him and could not reach him by telephone I tried to look him up at home, where I found his mother, who was worried and upset. She gave me his address out of town and asked me to visit him. I found him slowly convalescing. When I saw him again in

1936, he told me that he was striving to prove the consistency of the continuum hypothesis in set theory.

9.

At the time of Hitler's coming to power in Germany, there were periods of almost unbearable tension in Vienna.[27] Gödel kept himself well informed and spoke with me a great deal about politics[28] without showing strong emotional concern about the events. His political statements were always noncommittal and usually ended with the words "don't you think ?"

For a short time, things became somewhat calm; but then came the two civil wars of the year 1934. After the second, which began with the murder of Chancellor Dollfuss by the Nazis, the situation soon stabilized — outwardly — under the immense pressure of the new Schuschnigg regime. But even in the face of the seemingly inescapable dilemma between a Europe ruled by Hitler and a second world war, Gödel remained rather impassive. Occasionally, however, he made pointed and original remarks about the situation. Once he said to me: "Hitler's sole difficulty with Austria lies in the fact that he can only appropriate the country as a whole. Were it possible to proceed piecewise, he would certainly have done so long ago — don't you think?"

It was a tragic spectacle to observe the atrophy of the previously vigorous intellectual life in Vienna outside of the university — a consequence of propaganda, pressure and fear. The extreme nationalists ruled in the faculty as well as in the student body in the university. The Schlick Circle, which already had a world reputation, was disparaged and maligned, the Colloquium, one of the most active centers of mathematical research, was totally ignored, Gödel and the new logic were never mentioned. Objective arguments were considered irrelevant, especially since Hahn, a tireless and effective speaker for progressive causes, had died in 1934. Schlick remained one of the few defenders of reason at the university. For liberal-minded people who had loved Austria, life in Vienna became harder with every passing month.

Then, in June, 1936, Schlick was shot dead in a university building by a psychopathic former student of his. Still under the impact of the tragedy, I described the situation to some friends at the International Congress of Mathematics in Oslo. A few weeks later, I received a cable offering me a professorship at the University of Notre Dame in Indiana. I found it very hard to give up the Colloquium, but in Vienna the time for meetings of the type of ours was obviously running out, while I hoped that I might have the opportunity to develop a similar group in America. And it was very sad to leave Gödel, Wald, and other others, but I trusted that before long I should see them in the free world. In January, 1937, I took up my new position.

10.

The University of Notre Dame in South Bend, Indiana, about 90 miles east of Chicago, was founded in 1842 by a Catholic order (originally French). In the 1930's President O'Hara, an extraordinarily energetic and resourceful priest, invited scholars of many countries to his school. Among the élite of the religious order I found very cultivated men motivated by unselfish ideals, and I resolved to help them develop advanced studies of mathematics at the University.

In the meantime the European situation, and especially that of Austria, became darker and darker, and I was quite concerned about

Gödel. In particular I feared that, despite his cautiousness, an injudicious or misunderstood remark of his might have incalculable consequences. At the same time, my efforts to put Notre Dame on the mathematical map were bearing fruit as evidenced by the heavy attendance at a demanding mathematical symposium during the spring semester of 1937.[29] In the afterglow of this, I decided to speak to Father O'Hara about Gödel and his discoveries.

At the Catholic universities in the United States in the mid-1930's, logic was completely dominated by the writings of Jacques Maritain and the philosophical school of Laval University in Quebec, both quite opposed to mathematical ('merely formal') logic. In addition, there was the antipathy of religious circles for Bertrand Russell, with whom many almost identified mathematical logic. Nonetheless, already in that first conversation (in which I also mentioned that Gödel was a Protestant) President O'Hara showed great interest in Gödel's work and had literature on Notre Dame sent to his Vienna address. After a second conversation, he asked me to inquire whether Gödel would care to visit the school. This I communicated to Gödel, whereupon he wrote from Vienna on July 3, 1937 as follows:

". . . I am agreeable in principle to coming to the University of Notre Dame; indeed it would interest me very much to get to know the workings of an American Catholic university. The bulletins sent to me interested me very much and I am very grateful for them. . . For me, the summer semester of 1938 would be the earliest possible date to be considered . . . It is essential for me to be under obligation for *one* semester only. As you know, I have had bad experiences with my health in America, and hence do not want to bind myself in advance for a longer period. But if the conditions mentioned are met, I would gladly accept. . .

"There is not much new with me. Since I returned from Aflenz[30] my health is poor, though tolerable. Mr. Wald has probably written you about my lecture, at the end of which I proved the consistency of the axiom of choice for the system of set theory. The analogous proof for the system of *Principia Mathematica* as well as the partial result on the continuum hypothesis which I mentioned to you, I reported in the Colloquium.[31] At the moment I am considering whether I should lecture (in Vienna) on something introductory or on something advanced, or whether I should not lecture at all and use my time for my own work. In the second case, there is the danger that I won't have any students, as not enough background in formal logic or set theory is available now that Carnap and Hahn are no longer lecturing.

"In conclusion, I want to congratulate you heartily on your appointment at Notre Dame University, of which I learned from the bulletin, even though I also very much regret having lost another friend in Vienna. . ."

At the same time, Gödel wrote to President O'Hara, who, however, never received the letter. In early November, 1937, I urged Gödel to let me know his decision concerning Notre Dame, after which he wrote me from Vienna on December 15, 1937:

"I have decided not to come to America after all in the current academic year, as I already definitively wrote in the lost letter to President O'Hara. At the moment I cannot even definitely accept an offer for 1938-39, but I will be able to write you about that in about two months. In any event I would be very happy to spend another semester with you. If any part of the Symposium on the Algebra of Geometry[32] is printed, please send me offprints. I have unfortunately not received the program of your Symposium on the Calculus of

Variations which you mentioned. That also would interest me very much.

"I continued my work on the continuum problem last summer and I finally succeeded in proving the consistency of the continuum hypothesis (even the generalized form, $2^{\aleph_a} = \aleph_{a+1}$) with respect to general set theory. But I must ask you for the time being not to tell anyone about this. So far, apart from yourself I have communicated this only to von Neumann, for whom I sketched the proof during his latest stay in Vienna. Right now I am trying also to prove the independence of the continuum hypothesis, but do not yet know whether I will succeed with it . . ."

In the spring of 1938, Gödel decided to spend the year 1938/39 in America. When by the end of May I had not yet heard about his plans for the fall of 1938, I had to write to him saying that it was too late for the announcement of lectures in September. To this he answered on June 25, 1938 (three months after the occupation of Austria by Hitler's troops):

". . . In the meantime, I have abandoned the plan to come to Notre Dame in the fall, because it would after all be too strenuous for me. So I'll come in February, 1939, unless unforeseen circumstances should prevent me.

"As to the problem of lectures, I believe that at present I am not quite up to offering an introductory course because of inadequate English, lack of experience in introductory teaching and insufficient time for preparation. On the other hand, in a seminar on the continuum problem of only two hours per week I would have to omit much that is of interest and also to place high demands on the students. Hence I would prefer to give a 3-hour course on the axiomatization of set theory, in which I would take up results of myself and others on the continuum hypothesis and the axiom of choice and their relations to cardinal arithmetic. I would presuppose only knowledge of abstract set theory of about the extent of the first chapters of Hausdorff *Mengenlehre* (2nd edition). The more elementary portions of formal logic I would have to supply in any case."

Gödel wrote me again on October 19, 1938:

". . . I have been here in Princeton since October 15, and am thinking of staying till the end of December. I hope that you received my letter of late June, in which I wrote that I would like to spend the spring semester of 1939 in Notre Dame . . . From Veblen I hear you will be at the Meeting of the Mathematical Society in New York on October 29. I too am planning to come and would be extremely glad if I could meet you there. Then we could also discuss all the details concerning my lectures at Notre Dame. As I wrote before, in my last letter, I fear that an introductory course may not succeed for the reasons mentioned. But if you believe that I am wrong about this, and if the university lays great value on it, then I am also prepared to take up the proposed plan. . . "

On November 11, 1938, he wrote me about the matter a last time, from Princeton.

"I have once more thought over the question of my courses at Notre Dame and wish to say the following: If you prefer to announce a joint seminar, I am perfectly in agreement with that too under the condition that only those should give talks who have sufficient command of the subject, and secondly, that the introduction to logic, which will have to occupy a large part of the semester, be done somehow or other in a uniform manner (e.g., that only I, or you and I together, give the required talks). Afterwards one could perhaps insert reports by others on somewhat more difficult problems. As I said, I leave the decision to you, but perhaps a seminar would be preferable for the reasons mentioned in my two last letters . . . "

11.

After this long correspondence, which testifies to Gödel's conscientiousness and his desire to be of use, he gave a course in logic at Notre Dame in the spring of 1938. He described it later in a letter to me by saying that his 1940 Princeton lectures "contained the same in somewhat better form." For the beginning of the course which was devoted to the usual propositional calculus, we distributed mimeographed notes, which Gödel then discussed in the lectures.

At the beginning, Gödel had an audience of about 20 — half of them mathematicians who attended his lectures to their conclusion. These were young instructors and doctoral candidates with good preparation in mathematics (if not specifically in mathematical logic). Among them was a group who were developing, under my guidance, a new axiomatic approach to the non-Euclidean geometry of Bolyai and Lobachevsky[33] and this seemed to interest Gödel. Once or twice, Emil Artin attended Gödel's lectures when he came every other week, as he did, from Indiana University to Notre Dame to give a course on algebra.

The other half of the audience consisted of older philosophers and logicians, occasionally joined by one or another member of the physics department. For the reasons mentioned in Section 11, the older logicians did not evince particular interest in Gödel's lectures. Of the philosophers, Professor Yves Simon, a student and friend of Maritain's whom President O'Hara brought over from France, made a special effort to take advantage of Gödel's presence.

In informal gatherings of young instructors and doctoral candidates, I was not successful in developing an atmosphere resembling that of the Vienna Colloquium. Nor was my hope that Gödel's presence would help to bring the meeting closer to the style of those of Vienna fulfilled. Moreover, I did not keep minutes. Hence the *Reports of a Mathematical Colloquium*, (of which altogether eight issues appeared in Notre Dame) do not resemble the main protocol-like part of the Vienna *Ergebnisse*, but rather the *Mitteilungen* (Communications) at the end of each Viennese issue: i.e. articles by members and visitors relating to subjects discussed in the Colloquium.

During his stay at Notre Dame, Gödel appeared to be in fairly good health, but not particularly happy. He lived on campus — I do not recall whether for the whole period, but certainly for a large part of the semester. But he had quarrels with the prefect of his building for various trivial reasons (because of keys and the like). I found it not always easy to settle the differences, since the prefect was an old priest, set in his ways, while Gödel was emphasizing his rights.

12.

Gödel esteemed some Scholastic logicians as being forerunners of George Boole and Augustus de Morgan; and he asked Simon which earlier logicians especially interested *him*. Simon recommended the writings of John of St. Thomas, a Portuguese Thomist of the first half of the seventeenth century. Gödel looked at these writings (with me), but did not find anything of special interest to him.

I have been asked whether Gödel expressed interest in Catholicism. In my presence he only once touched on such a topic: he one day asked me, all of a sudden: "Is there actually a list of all the saints of the Catholic Church?" I referred him to a church historian. Whether he followed the matter up I do not know.

Gödel leaned more and more towards Platonism. In his *philosophical* thinking at that time two thoughts were important even though he published them only in the 1940's:[34] 1). The assumption of classes and the belief in the existence of classes are just as legitimate as the assumption of bodies and the belief in the existence of bodies. 2). Classes are just as necessary for a satisfactory system of mathematics as bodies are for a satisfactory theory of sensations and of physics. Ernst Mach had virtually *defined* bodies as certain classes of elements of our sensations, namely classes of elements that we always find together, though, in keeping with his antimetaphysical position, Mach of course eschewed assertions about the *existence* of bodies or about our *belief* in them. From this positivistic attitude, however, Gödel dissociated himself more and more articulately, emphasizing instead the pragmatic argument. 3). In Gödel's *mathematical* thinking, the most realistic element — realistic in contrast to nominalistic — was probably his early conviction that the right axioms of set theory had not yet been discovered (Cf. Section 6). He undoubtedly meant that no one had given an adequate basic description of that world of sets in which he believed — a description that would permit us to decide the fundamental problems of cardinality such as Cantor's continuum hypothesis.

This conviction he expressed more and more emphatically, but I myself never heard from him any indications about where he expected to find such axioms.

Meanwhile, Gödel was more and more preoccupied with Leibniz. He was now completely convinced that important writings of this philosopher had not only failed to be published, but were destroyed in manuscript. Once I said to him teasingly, "You have a vicarious persecution complex on Leibniz' behalf." Soon afterwards he said, "There is something I have wanted to ask you for quite a while. When was the Viennese (now Austrian) Academy of Sciences founded?" I immediately suspected what Gödel was after. It is a historical fact that Leibniz negotiated for a time with the Emperor and his government about the founding of an Academy in Vienna, but that the negotiations came to nothing. My answer to Gödel's question was, "In the year 1846, under the predecessor of Emperor Franz Josef." Gödel was visibly disappointed and replied: "You are saying what everyone else says." "What kind of answer did you expect from me?" I asked. "At the time of Leibniz, of course!" he said. "In the Proceedings of the Viennese Academy, there appeared important writings of Leibniz which, however, were destroyed." I reminded him of the stranded negotiations and asked him: "How could the founding of the Academy be kept secret for centuries? How could its Proceedings disappear without a trace? Who had an *interest* in destroying Leibniz' writings?" "Naturally those people who do not want man to become more intelligent," he replied. Since it was unclear to me whom he suspected, I asked after groping for a response. "Don't you think that they would sooner have destroyed Voltaire's writings?" Gödel's astonishing answer was: "Who ever became more intelligent by reading the writings of Voltaire?" Unfortunately at that moment someone stepped into the room and the conversation was never concluded.

Later, I once discussed Gödel's ideas on Leibniz with a common friend, the economist Oskar Morgenstern. He described to me how Gödel one day took him into the Princeton University Library and piled up two stacks of publications: on one side, books and articles that appeared during or shortly after Leibniz' lifetime and contained exact references to writings of the philosopher published in collections or series (with places and years of publication, volume and page numbers, etc.); on the other side, those very collections or series. But in some cases, neither on the cited page nor elsewhere was there any writing by Leibniz; in other cases, the series broke off just before the cited volume or the volume ended before the cited page; in still other cases, the volumes containing the cited writings never appeared. "The material was really highly astonishing," Morgenstern said.

13.

While Gödel was at Notre Dame, Czechoslovakia was totally occupied by German troops and Hitler immediately afterwards prepared for war against Poland — events which greatly upset the Europeans (including myself) as well as many Americans at Notre Dame. In the second half of the semester, Gödel also, who until then had been his usual dispassionate self, appeared to be restless. Remarks of his indicated longing for his family. For this and other reasons he wanted to return to Vienna at the end of the semester. Even earlier he had complained about the revocation of his dozentship in the University by the Nazi regime and had spoken about violated rights. "How can one speak of rights in the present situation?" I asked, "And what practical value can even *rights* at the University of Vienna have for you under such circumstances?" But despite pleas and warning by all his acquaintances at Notre Dame and Princeton, he was determined to go to Vienna; and he went.

After a while, rumors reached me that Nazi rowdies had attacked him on the street and knocked his glasses off. The only reason was his obviously intellectual appearance and manner, which these fellows took for semitic — in Gödel's case, as in many others, erroneously. Later I learned that despite everything, he established his rights to the dozentship in Vienna.

During the summer I heard nothing from Gödel. But on August 30, 1939, one of the few days between the Hitler-Stalin pact and the entry of German troops into Poland which unleashed the second world war, he wrote me a letter that may well represent a record for unconcern on the threshold of world-shaking events:

"Since the end of June I have been here in Vienna again and had a great many tasks to perform so it was unfortunately not possible to write up anything for the Colloquium. How did the examinations turn out for my logic lectures? . . . In the fall I hope to be back in Princeton. . ."

He let Veblen know that he wanted to return to America but that his exit was refused. Later Veblen told me that Dr. Abraham Flexner, the Director of the Institute for Advanced Study, worked for hours on a letter to the German government to obtain an exit visa for Gödel. The letter, which Veblen characterized as a masterwork of diplomacy, achieved the miracle of getting Gödel out of Germany even though he was subject to the draft. I was immensely relieved to know that he was safe again; but I must confess that it was not easy to find in me all the warmth that I used to feel for him.[35]

Soon after Gödel had left Notre Dame, plans began for the centenary celebration of the University in 1942. The contribution on the part of mathematics was to be a series of booklets Notre Dame Mathematical Lectures, beginning with Wald's Notre Dame lectures On the Principles of Statistical Inference and Artin's lectures on Galois Theory, taken, at my urging, from notes by Arthur N. Milgram. The series was to be continued with Gödel's Notre Dame lectures. But soon after the beginning of the European war, long before the entry of the United States, the Navy established at Notre Dame one of its largest training centers for aspiring Naval officers. The work load, especially for mathematicians, rose drastically — and, after the entry of the United States into the war, up to the limit of feasibility. On the other hand, a letter from Gödel indicated that the editing and publication of his lectures would require more work than could possibly be managed at Notre Dame at the time. The centenary celebration was indefinitely postponed and Gödel's lectures never appeared.

14.

After the war, I accepted an invitation to the newly founded Illinois Institute of Technology in Chicago. The chairman there, Dr. Lester R. Ford, had been among those at the Rice Institute who had heard of Gödel as early as 1931. As the editor of the *Mathematical Monthly*, he initiated a very interesting series of articles whose titles began with the words '*What is . . . ?*' His request for a contribution from Gödel gave rise to the article 'What is Cantor's Continuum Problem?'[36]

On every one of my admittedly infrequent visits to Princeton, I had long talks with Gödel. Apart from his friendship with Einstein and (especially after the latter's death) with Morgenstern, Gödel seemed to me rather lonely. Once he asked to my surprise, "Where is Artin now?", and when I answered, "In Princeton; I spoke to him yesterday," Gödel said, "I thought he left long ago. I haven't seen him for years."

At no time in his life did Gödel need intellectual stimulation to conceive and develop original and unexpected ideas. But he needed a congenial group suggesting that he *report* his discoveries, reminding and, if necessary, gently pressing him to write them down. All this he *had* at the beginning of his stay in Princeton with regard to the publication of his two booklets and his article on Russell. And he presumably could have found such support later. But apparently he never looked for it, and no one seemed to volunteer. The fact is that I could not observe anything of the sort in the 1950's. Rather, it soon became clear to me that he wrote up many brillant ideas only for his desk drawer if at all. From the point of view of the outside world, his incomparable talent was lying lamentably fallow. But it was clear that Veblen was determined to retain Gödel at Princeton, and spared no sacrifice to this end.

15.

In the 1950's my conversations with Gödel mainly revolved around some fundamental concepts of pure and applied analysis which I was examining critically at that time. In the first decade of this century, Russell had started an investigation of the concept of *variable*, which he called one of the most difficult in mathematics, but he soon abandoned these studies without having achieved a result that he found satisfactory.[37] Subsequently, the problem fell more or less into oblivion.

Originally I had planned to eliminate variables from analysis as much as possible. Gödel said, "It is as though you wanted to speak only about mankind, never simply about a man." I clarified the program. Number theory and pre-abstract algebra require, as has been clear since Vieta, number variables (letters that stand for any number or any number of a certain kind) in the formulation of comprehensible general assertions. Similarly, analysis needs function variables (letters that stand for any function or any function of a certain kind). But the omnipresence of the number variables x and y in analysis is a mere consequence of fundamental gaps in the symbolism of functions that has come down to us from the Renaissance: the lack of designations of the identity, the n-th power and the constant functions (e.g. of the values 3 or a) which are traditionally denoted by their values for x ('the functions x, x^n, 3, a'). For the sake of uniformity, this notation is extended even to the logarithmic, the cosine and other functions and to function variables, which are designated by *log*, *cos*, and *f*, while being called 'the functions log x, cos x, f(x)'. Filling the gaps by symbols such as *j*, *j^n*, c_3, c_a, (analogous to *log, cos, f*) *would result in an unambiguous, uniform notation for analysis* and is absolutely indispensable for the development of an *algebra of functions*, where the identity function plays a role similar to that of 0 and 1 in arithmetic.

As to the term *variable*, Gödel emphatically insisted on its exclusive use in the logico-mathematical sense of a letter that stands for any element of a certain class (as number variables and function variables have been defined above). It so happened that a few weeks after my talk with Gödel an eminent physicist equally emphatically insisted on the exclusive use of 'variable' in the scientific sense, i.e. for pressure, volume, temperature, time elapsed, distance travelled and the like, which Gödel referred to as *variable quantities* and which, for the sake of brevity, I proposed to call *fluents*. Over such terminological and hence totally unimportant differences, one often overlooks the profound *conceptual* difference between variables (letters that do not designate anything individual or specific) and fluents such as the ones mentioned above, which usually are denoted by single letters P, V, T, t, s (each of which designates a specific entity). Moreover, applied analysis operates with a third notion, not at all articulated before the mid-1950's. It is exemplified by u and w in the assertion

(*) $w = u^2$ implies $dw/du = 2u$.

In (*), u and w are neither number variables ('9 = 3^2 implies d9/d3 = 6' is nonsense) nor fluents (they do not designate any specific entities). They stand for any fluents of certain kinds, e.g. for t and s in the case of balls rolling on certain inclined planes or for side and area of expanding squares. Thus **u** and **w** are *fluent variables*.

The preceding exhibition of totally different concepts indiscriminately called 'variable' (against the background of Russell's abortive attempt to define 'the' concept of variable) must be supplemented by a definition of fluents. A *fluent* is a certain pairing of numbers to objects such as rolling balls or expanding squares. Is a fluent thus a function? As in the case of the notion of variables, terminological considerations must not obscure important conceptual differences. A function such as *log* or *cos* — being a certain class of pairs of numbers — is a purely *mathematical* concept connecting all kinds of fluents and hence usable in all branches of science. A fluent has a nonmathematical domain and, therefore, is an extramathematical concept, limited to a special part or aspect of the world, and of interest only to special branches of science. Moreover, there are only functions (but not fluents) of fluents and of functions. There is a logarithm of the cosine and of the temperature but no temperature of the logarithm or of the pressure.

The conceptual clarification in the preceding paragraphs seem to me to be indispensable for an articulation of the rules for the application of analysis to science — as it were, for the *theory of application*.[38]

Upon receipt of my early publications on the subject, Gödel wrote me in 1953 "I believe one should beware that the explicated (präzisierte) meaning of a word be never *completely* different from the meaning the word previously had, and further that one never designate *completely* different things with the same word." For my part, I have always been of the opinion that the preceding conceptual clarification implemented the very idea expressed in Gödel's letter in which I had always believed. But Gödel, I am sorry to say, never confirmed this opinion.[39]

16.

In the mid 1930's I had inferred from some remarks of Waismann's in the Schlick Circle that Wittgenstein did not know of Gödel's results on decidability and consistency. I felt that Waismann was to blame for not informing the philosopher about those fundamental ideas. In 1967, when Wittgenstein's conversations with Waismann in the years 1929-32 were published[40] I found my suspicion confirmed. Consistency was repeatedly a topic of discussion, but merely from a pre-Gödel standpoint.

In the early 1970's I began writing a book on my recollections of the Schlick Circle. For the sake of completeness, I looked for ideas about Gödel published by Wittgenstein. A few were in the latter's book *Remarks on the Foundations of Mathematics*, which appeared in 1956. Aside from noncommittal remarks in Part 5, the Appendix I of Part I (written in 1937-38) contains a discussion of the problem — without, however, any adequate appreciation of Gödel's work. In fact, Wittgenstein goes so far astray as to say that the only use of undecidability proofs is for '*logische Kunststückchen*' (little logical artifices or tricks).[41] After a close reading of that Appendix, I asked myself whether Wittgenstein had completely understood Gödel's results. To make quite sure, I presented my question to Gödel and also asked him whether he had ever spoken to Wittgenstein. On May 20, 1972, Gödel wrote me that he had seen Wittgenstein only once in his life (and on that occasion only from a distance), and then he answered my main question as follows:

"As far as my theorems about undecidable propositions are concerned, it is indeed clear from the passage that you cite that Wittgenstein did *not* (Gödel's emphasis) understand it (or that he pretended not to understand it). He interprets it as a kind of logical paradox, while in fact it is just the opposite, namely a mathematical theorem within an absolutely uncontroversial part of mathematics (finitary number theory or combinatorics). Incidentally, the whole passage you cite seems nonsense to me. See e.g. the 'superstitious fear of mathematicians of contradictions'."

17.

In Gödel's last years we were again in more or less constant contact, but only indirectly, namely through Oskar Morgenstern. Although the latter was himself gravely ill, he telephoned from time to time, always transmitting greetings from Gödel (and on two occasions invitations, which I unfortunately had to decline) and speaking a great deal about our friend, especially about his diverse illnesses and difficulties with local physicians. I asked Oskar to relay to Gödel that, whatever his complaints, he should find out the name of the best New York specialist and then travel the 40 miles or so from Princeton to New York for an examination and, if necessary, for treatment. In one case Morgenstern reported that Gödel followed this obvious advice.

In one of his last telephone calls before his own death (in July, 1977), Morgenstern described an event that evoked in me memories that long ago had somewhat estranged me from Gödel — but it evoked them by its *contrast* to those memories, so that Morgenstern's story moved me very much. Once again it was a question of Gödel's rights, where his punctiliousness knew no bounds. What had happened was that Gödel, apparently suffering severely, sought and was granted admission to a Princeton hospital, but soon thereafter insisted that he had no right to one of the benefits proffered, since his insurance policy did not provide for it. He therefore refused to accept the benefit. The details of the case escape me now, though of course I am convinced that Gödel's logic in interpreting the insurance contract was superior to the hospital's. But be that as it may, in his juridical precision, Gödel unshakably maintained his ground, even though — according to Morgenstern — the hospital routine was disturbed, inconvenience arose on all sides and, what was the most grievous result, Gödel was deprived of some paramedical help.

He died in January, 1978.

*"The translation of this chapter, which we have lightly edited, is the work of Eckehart Koehler, with some contributions by Karl Menger. We thank Mr. Koehler for the opportunity to use it here, in accordance with Menger's wishes."

Notes

[1] *Jahresbericht d. deutschen Mathematikervereinigung* 37 (1928) 213-226, and 298-308. Cf. also my book *Selected Papers on Logic and Foundations, Didactics, Economics*, Dordrecht 1979, quoted subsequently as *Selected Papers*, Chapter V.

[2] For the development of the logical Principle of Tolerance cf. *Selected Papers*, pp. 11-16.

[3] *Erkenntnis* 2, (1931), 147f.

[4] *Ergebnisse eines mathematischen Kolloquiums*, 3 (1931-32) 12f. In the fall of 1930, a summary of Gödel's fundamental results appeared in the *Anzeiger* of the Austrian Academy of Science, Nr. 19. His famous paper 'Über formal unentscheidbare Sätze der *Principia mathematica* und verwandter Systeme, I' (On formally undecidable

propositions of *Principia mathematica* and related systems I) was published in *Monatshefte für Mathematik*, 38 (1931) 173-198, (also Kurt Gödel *Collected Works* Oxford and New York, 1986, hereafter *CW*, pp. 141-195) see also the short summary in *Erkenntnis 2*, 1931 149f.

[5]*Ergebnisse 3*, 20f. (*CW*, pp. 238-41) Without any appeal to *true* and *false* the formalism of the traditional propositional calculus can be based on the following assumption: Let $\mathfrak{S}$ be the set of undefined elements (called propositions) with:

1) two operations: one unitary, denoted by $\sim$ (read *not*), and one binary denoted by $\supset$ (read *implies*);

2) a nonempty subset $\mathfrak{T}$ (whose elements are called *tautologies*) satisfying the following conditions:

α) if p and p $\supset$ q belong to $\mathfrak{T}$, then so does q;

β) the axioms of the traditional propositional calculus, e.g. those of Lukasiewicz:
$(\sim p \supset p) \supset p, \quad p \supset (\sim p \supset q), \quad (p \supset q) \supset [(q \supset r) \supset (p \supset r)]$
are elements of $\mathfrak{T}$.

Under this assumption, Gödel proves that $\mathfrak{S}$ is the union of two disjoint subsets $\mathfrak{W}$ and $\mathfrak{F}$. (whose elements may be called *true* and *false*, repectively), with the following properties:

a) for each pair of propositions p, $\sim$ p, one belongs to $\mathfrak{W}$ and the other to $\mathfrak{F}$:

b) the proposition p $\supset$ q belongs to $\mathfrak{F}$ if and only if p belongs to $\mathfrak{W}$ and q to $\mathfrak{F}$.

Calling a set $\mathfrak{S}$ with two operators, $\sim$ and $\supset$, and a subset $\mathfrak{T}$ satisfying the assumptions 1) and 2) a *model* of the traditional propositional calculus one can formulate the preceding theorem as follows: *The elements of each model of the traditional calculus of propositions fall into two classes whose elements behave exactly like the true and the false propositions of that calculus.*

I should like to add here that my original question to Gödel may be extended to multi-valued propositional calculi.

[6]'Die neue Logik' in the booklet quoted in 7 pp. 94-122. Translated as 'The New Logic' in *Philosophy of Science* 4, 299-336, and reprinted in *Selected Papers*, pp 17-45.

[7]*Krise und Neuaufbau in den exakten Wissenschaften*, Leipzig und Vienna, 1933, 122pp.

[8]*Review of Scientific Instruments* 4 (1933). Cf. *Selected Papers*, p. 17f.

[9]Gödel mentions three propositions satisfied in each model such that $p \supset \sim \sim p$, while there exists an infinite model not satisfying them.

[10]C. I. Lewis' system had been extensively discussed by Dr. W. T. Parry (4,15f). In a discussion after an earlier paper by Dr. Parry, Gödel had proposed the following interpretation of Parry's 'p implies q analytically': 'q can be derived from (*ist ableitbar aus*) p and the axioms of logic, and contains no other concepts than p' (4,6, CW pp. 266-7).

[11]*Math. Annalen* 100 (1928) 77f.

[12]*Math. Annalen* 104 (1931) 476-484.

[13]The W is characterized among the subsets of (*) by the following properties: If the point (x,y,z) belongs to W, then so does (y,x,z), but not (z,y,x). If (u,v,w) and (w,x,y) belong to W, then so do the points (u,x,y) and (v,x,z) for every z provided that they belong to (*). For each z >0, the set W contains at least one point (x,y,z). The set consisting of (0,z,z), (z,0,z) and all points (x,y,z) in W is closed.

[14]*Jahresber. DMV* 37 (1928), *Annals of Math.* 37 (1936) 456-482.

[15]Cf. *Ergebnisse* 1, 28f.; 4,14f, and 34,7; 11f, and *Annals of Math.* loc. cit. 14.

[16]*Math. Annalen* 103 (1930) 466-501.

[17]The radius of the circumcircle of three points is defined metrically as the quotient of the product of the three sides by the area of the triangle (expressed by Hero's formula).

[18]'An example of a new type of cosmological solutions of Einstein's field equations of gravitation.' *Review of Modern Physics* 21, 1949, pp. 555-62.

[19]*Wissenschaftliche Weltauffassung: Der Wiener Kreis*, Wien 1929. E. T. in O. Neurath, *Empiricism and Sociology*, Vienna Circle Collection, Dordrecht.

[20] *Elements d'Economie Politique Pure*, 2 ed. Lausanne 1889.

[21]*Ergebnisse* 6 10 f.

[22]*Ergebnisse* 6, 12-18.

[23]*Ergebnisse* 7, 1-6.

[24]*Ergebnisse* 7, 6.

[25]Earlier that year, Gödel participated in a discussion by Tarski and myself on

Friedrich Waismann's *Bemerkungen zu Freges und Russells Definition der Zahl*, (Remarks on Frege's and Russell's definition of numbers) no details of which have been recorded. [Waismann's paper was no doubt identical in content with the relevant section of his *Einführung in das Mathemathische Denken* (E. T. *Introduction to Mathematical Thinking*) Vienna 1936 (New York, 1951) B. McG.]

[26]*Ergebnisse 7*, 23, *CW* pp. 396-399.

[27]About that period, cf. my 'Memories of Moritz Schlick' in *Rationality and Science* ed. E. T. Gadol, Vienna 1982.

[28]I do not recall any instance of his discussing socio-economic questions.

[29]A three-day meeting on all aspects of the calculus of variations.

[30]A resort in Styria.

[31]Alt and Wald arranged a couple of meetings after my departure, as they had promised me; but no minutes were kept.

[32]A three day meeting on algebra of geometry and other aspects of lattice theory. No proceedings were published.

[33]Cf. papers by F. P. Jenks, J.C. Abbott, B. T. Topel, J. Landin in the first 6 issues of *Reports of a Mathematical Colloquium, 2nd Series*, Notre Dame, 1938-45.

[34]'Russell's mathematical logic' in *The Philosophy of Bertrand Russell, Library of Living Philosophers*, ed. P. A. Schilpp, Evanston 1944, 123-153.

[35][Remark by Eckehart Köhler: In June 1985, shortly before his death in October 1985, Karl Menger expressed to me in a telephone call that he had only recently learned of the fact that Gödel had married Adele in September 1938, and that had he known that in 1939-40, his attitude to Gödel's return then and subsequently would have been completely different; and that the corresponding passages of this memoir on Gödel should be amended.]

[36]*Amer. Math. Monthly*, 54 (1947) 515-525.

[37]*Principles of Mathematics*, Cambridge 1903, p. 5 and p. 89.

[38]All these matters are extensively treated in my *Selected Papers*, Parts III and IV. The clarifications are of course also relevant for the didactics of pure and applied analysis and of mathematical science. But this is a field in which Gödel (in contrast to his occasional interest in logical didactics) had no experience and took only little interest.

[39] On the other hand, J. von Neumann told me (and our common friend Morgenstern) that after thinking over the arguments presented above he became aware of previously unnoticed obscurities in the mode of expression in our mathematico-scientific discourse.

[40] Friedrich Waismann, *Wittgenstein und der Wiener Kreis*, ed. B. McGuinness, Oxford 1967. (E.T. *Wittgenstein and the Vienna Circle*, Oxford, 1979)

[41] [In all subsequent editions this is the third appendix to Part I: the passage referred to occurs at the end of the penultimate paragraph. The word "logische"/"logical" does not occur in Wittgenstein's text, though the word "Kunststücken"/"conjuring tricks" or "legerdomain" does. B. McG.]

INDEX OF NAMES

Vienna Circle Collection

1. Otto Neurath: *Empiricism and Sociology*. With a Selection of Biographical and Autobiographical Sketches. Translated from German by Paul Foulkes and Marie Neurath. Edited by M. Neurath and R.S. Cohen. 1973
ISBN 90-277-0258-6; Pb 90-277-0259-4

2. Josef Schächter: *Prolegomena to a Critical Grammar.* Translated from German by Paul Foulkes. With a Foreword by J.F. Staal and the Introduction to the Original Edition by M. Schlick. 1973   ISBN 90-277-0296-9; Pb 90-277-0301-9

3. Ernst Mach: *Knowledge and Error. Sketches on the Psychology of Enquiry.* Translated from German by Thomas J. McCormack and Paul Foulkes. With an Introduction by Erwin N. Hiebert. 1976 ISBN 90-277-0281-0; Pb 90-277-0282-9

4. Hans Reichenbach: *Selected Writings, 1909-1953.* With a Selection of Biographical and Autobiographical Sketches. Translated from German by Elizabeth Hughes Schneewind and others. Edited by M. Reichenbach and R.S. Cohen. 1978, 2 vols.   Set ISBN 90-277-0892-4; Pb 90-277-0893-2

5. Ludwig Boltzmann: *Theoretical Physics and Philosophical Problems. Selected Writings.* Translated by Paul Foulkes. Edited by B. McGuinness. With a Foreword by S.R. de Groot. 1974   ISBN 90-277-0249-7; Pb 90-277-0250-0

6. Karl Menger: *Morality, Decision and Social Organization. Toward a Logic of Ethics.* Translated from German by Eric van der Schalie. 1974
ISBN 90-277-0318-3; Pb 90-277-0319-1

7. Béla Juhos: *Selected Papers on Epistemology and Physics.* Translated from German by Paul Foulkes. Edited and with an Introduction by Gerhard Frey. 1976
ISBN 90-277-0686-7; Pb 90-277-0687-5

8. Friedrich Waismann: *Philosophical Papers.* Translated from German and Dutch by Hans Kaal, Arnold Burms and Philippe van Parys. Edited by B. McGuinness. With an Introduction by Anthony Quinton. 1977
ISBN 90-277-0712-X; Pb 90-277-0713-8

9. Felix Kaufmann: *The Infinite in Mathematics. Logico-mathematical Writings.* Translated from German by Paul Foulkes. Edited by B. McGuinness. With an Introduction by Ernest Nagel. 1978   ISBN 90-277-0847-9; Pb 90-277-0848-7

10. Karl Menger: *Selected Papers in Logic and Foundations, Didactics, Economics.* Translated from German. 1979   ISBN 90-277-0320-5; Pb 90-277-0321-3

11. Moritz Schlick: *Philosophical Papers.*
Vol. I: *1909-1922.* Translated from German by Peter Heath, Henry L. Brose and Albert E. Blumberg. With a Memoir by Herbert Feigl (1938).
Vol. II: *1925-1936.* Translated from German and French by Peter Heath, Wilfred Sellars, Herbert Feigl and May Brodbeck.
Edited by Henk L. Mulder and Barbara F.B. van de Velde-Schlick. 1979
Vol. I: ISBN 90-277-0314-0; Pb 90-277-0315-9
Vol.II: ISBN 90-277-0941-6; Pb 90-277-0942-4

Vienna Circle Collection

KLUWER ACADEMIC PUBLISHERS – DORDRECHT / BOSTON / LONDON